Rules for Eternity

Author: John Hunter

www.rulesforeternity.com

Copyright © 2015 - 2025 John Hunter
First printed edition 2024
All rights reserved.
ISBN: 978-0-9942620-5-9

DEDICATION

To the loving parents who created the circumstances that permitted me to create.

J Hunter

CONTENTS

Table of Contents

Preface...i
Acknowledgement..iii
Introduction..v
1. Endless Constraints on Existence............................1
2. Biology – The Rules for Life..................................21
3. The Human Factors..35
4. Why We Need Gods..51
5. Universal Rules and Properties..............................63
6. Immutability and Instability.................................79
7. Science Fiction – A Reality Check..........................93
8. A Big Ask for Future Science................................109
9. Applying Science: The Challenges.........................125
10. What if the Rules Were Different.........................137
11. Big Bang and Before..155
12. No Beginning and No End...................................169
Reference Material..180
About The Author...185

PREFACE

The desire to seek answers to the really big questions is not restricted to Philosophers, Scientists, Poets, or Dreamers. The prospect of never answering many of these questions does not detract from the joy of the journey while researching the possibilities. In Rules for Eternity, we are not searching for, or offering, ultimate answers. The aim is to identify a common thread running through all aspects of the natural Universe, and to reflect on how this knowledge might have practical application to human civilization.

Science and Philosophy record Humankind's imaginative interaction with its environment in the broadest terms. As history is written, our ideas and beliefs are subject to revision and verification. Wherever we happen to be sitting on the evolutionary time-line, we are always at the boundary between the known and the unknown. The future provides increasingly fertile ground for theoretical physicists to explore, but the demands they place on engineering for proof are becoming prohibitive. Eventually Science will reach the limits of verification at both the sub-atomic and outer cosmic level, where none of the competing theories can be deemed more acceptable than the rest.

The last several hundred years have presented the perfect environment for individuals of astounding intellect to create watershed moments in science. There was a time when the Law-of-Gravity could be postulated because of an apple that fell on the head of a great mathematician. It now seems as though the 'easy' science has been done. To take our understanding to the next level requires an incredible investment in time, money, manpower, and cooperation. A perfect example of this is the Large Hadron Collider, which has so far yielded intriguing extensions to what was previously a simple concept involving gravitational-attraction between bodies with mass.

Rules for Eternity attempts to marry the science of the day with the unanswerable questions. Without wishing to spoil the plot, this book hopes to provide a convincing argument for what is stated on the front cover – "There is no beginning; There is no end". This presents a daunting challenge given our predisposition to think in terms of beginnings and endings, even when musing about Eternity.

ACKNOWLEDGEMENT

My most sincere thanks to Dr. Philip Rayment, a now retired mathematics and statistics academic. Without his enthusiastic encouragement and support during a lengthy gestation process, this book would never have become a reality.

Along with Philip's skill as a wordsmith honed over many years of assisting postgraduate students with the preparation of theses and publications, his 'attention to detail' has contributed greatly to the quality of Rules for Eternity.

However, it was not just Philip's skill with words that made him the ideal person to provide editorial guidance for this book. Some sections go into details that require a little background in basic physics. Philip was well qualified, not only to understand the physics at a basic level, but also to challenge some of the more esoteric concepts discussed.

He and I both fondly recall the inspirational lectures in cosmology delivered as part of university physics in Melbourne several decades ago.

John Hunter.

INTRODUCTION

Every aspect of our existence is controlled by Rules or Laws. Some of these Rules can be broken, with varying degrees of consequence. Invariably, these transient Rules are man-made. They can relate to a Continent, a State, or even attempt to define the minute constraints on where we might walk, sit, or smoke.

There are other Rules that must be obeyed without question, and there is not an opportunity to manoeuvre outside their restrictions, no matter how serious the intent. These unbreakable or Immutable Rules come from a higher authority, and are very much the focus of this book.

Many grey areas exist due to incomplete knowledge and understanding of the immutable rules. Identifying the exact scope of these rules will have a profound influence on the future directions available to Science. In the past, science-fiction writers have done a wonderful job of forecasting many advances in many fields of endeavour. Perhaps there is no better example than Arthur C. Clark, who first defined the Geostationary-Orbit, and clearly saw the place it would have in our technological future (ref: **Geostationary Orbit**).

But, in some cases, these same writers may possibly have done a disservice in creating a bar that might be too high to reach. There is a growing expectation that, given enough time, humans can do anything. While that attitude is to be admired, is it realistic? The purpose to this exploration of the Rules is hopefully to provide a reality-check on what is achievable, and what is pure fiction.

A requirement for seeing into the future is a solid understanding of the current rapidly expanding knowledge-base. It would be an utterly false expectation if we imagined we would ever be allowed or enabled to see the whole picture. Our entire world is but a pixel on the postage-stamp applied to an infinite universe. If we managed to eventually blow planet Earth back to a chaos of atoms and molecules, the effect would be less than a leaf-shake in the forest of our nearest galactic neighbours.

To examine the Rules to their fullest extent will require extrapolation way beyond our observable world. Once outside the observable event-horizon, we are squarely in the realms of Philosophy. In this book, we do go there, and make no pretence otherwise.

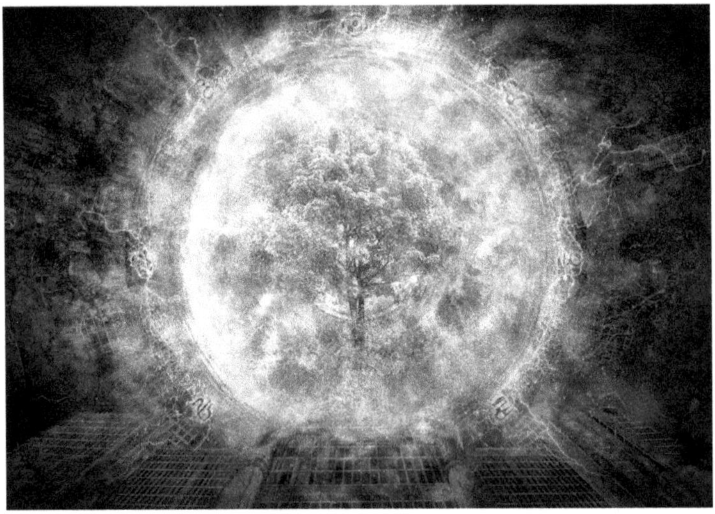

1. ENDLESS CONSTRAINTS ON EXISTENCE

As humans, every microsecond of our existence is governed by rules. At the lighter end of the spectrum, we create many of these rules ourselves, supposedly for the purpose of aiding the smooth running of society. At the other end of the spectrum sits a multitude of rules where we have no influence whatsoever. We are purely observers in this space, and one never-ending objective is to try to define the function and purpose for the rules of Nature. The intent of this book is to examine, perhaps philosophically, the way the natural rules interact with each other and ourselves.

In the broadest sense, Rules are either a description of, or a prescription for behaviour. As the Laws of Nature cannot be altered, scientific endeavour is confined solely to observing the results of the application of these laws, and devising endless experiments to test various hypotheses and models in an attempt to refine our understanding of the rules. In the case of man-made rules, they are meant to provide guidance for some desired behaviour. For these, there is often a voluminous written-record in the form of statutes and regulations. Rules-of-Nature are not available for reference in some earthbound library, and it is exceedingly difficult to imagine where the rules might actually reside. Although this presents a daunting challenge, there is something quite useful that does arise from exploring the thought. Much of the quest for mankind involves an investigation to discover origins. We have landed in a small slice of an infinite timeline. The 'origin of everything' occurred well outside anything that the human mind might imagine. Similarly, it is impossible to wrap the 'end of everything' in a way that has even a remote connection with our reality.

Much of the current human generation is immersed in the fruits of technology to an unprecedented extent. This can lead to a dangerous situation where we become collectively cocky about our place in the scheme of things. Because of the wonders technology has delivered thus far, we might be harbouring a belief that we could have complete control of our own destiny. However, it would be a fundamental mistake to ascribe to humans the power reserved for the gods. It is an interesting exercise to investigate how there will always be external constraints on human civilization. There is a need to ensure that we continue to play within the goalposts, and avoid spending fruitless effort in trying to change their location. This cockiness seems to be translating to a disrespect for our planet and a blasé attitude towards our long-term survival. It is hoped this book can bring some realistic perspective regarding our role in the giant scenario of life.

Firstly, a brief diversion into the history of man-made or transient rules, and a suggestion as to why these have become so prolific in modern society. Sometime over the last several million years, mankind developed the characteristics that set us apart from our near-relatives in the animal world. During the early stages in the evolution of man, virtually all rules were provided and controlled by Nature. As our civilizations advanced, the need arose for a more complex set of rules to accommodate the increasing stresses and demands of society. Prior to this, our ancestors were driven purely by natural rules or instinct. These natural rules were closely related to survival – the necessity to eat and drink; the drive to find a partner and procreate; the honing of skills required to kill anything that was trying to kill them.

A human in isolation would be driven purely by instinct. As he starts to form part of a larger collective, a set of rules develops that defines an appropriate moral code for the group. Often, any disregard for moral-rules is classified as a Sin. In many cases, the rules of morality are just a detailed version of what we consider to be instinct. However, quite different moral codes can exist for different societies or tribes, even within the same animal species. For most human societies, killing of one's own species is not considered acceptable. In the animal kingdom, this behaviour might sometimes be not only acceptable practice, but may be an essential part of the life-cycle for that species. There is no moral code among animals, only survival instincts, although the distinction between the two is not always clear-cut. To illustrate this point, there are many reported cases where conditions that threaten survival can turn animals into killers of their own species (ref: **Animal Infanticide**).

Ancient Egyptian and Greek civilizations were no strangers to rules. Both maintained a basic set of man-made rules, all created by the RULERS – the clue is in their title. However, there was a twist. Fearing man-made rules may not have sufficient authority to impress the masses, many of the rules were packaged as dictates from the gods. This had the added benefit that no particular ruler would need to take responsibility for any rule that might have had negative consequences. Even though there was a lack of hard evidence for the existence of gods, it seemed justifiable to allow for the possibility as good insurance. The gods of the day were not just about making rules on how society should run. They also provided a convenient explanation for the many things that were not understood. This practice continues today where we tend to explain the unknown as work of the gods (ref: **God of the Gaps**).

The basic tenet behind the creation of man-made rules was the assumption that they were essential to ensure the smooth functioning of society. That assumption is, in all likelihood, a valid one. In ancient times, there was little understanding of the natural rules that were already in place, guiding every step to maintain the health of the planet and to seek the best long-term outcomes for its inhabitants. In modern times, an important attempt to formally explain this process has been the Gaia Principle (ref: **The Gaia Hypothesis**). More and more we are discovering a need for global agreement, and rules, to ensure that collective human behaviour does not disrupt our planet's equilibrium. Currently, the obvious example is out-of-cycle warming of the Earth, caused by human population growth and industrialization.

The number of rules of society, or transient rules, has grown out of all proportion in modern times. The assumption is that every step must be controlled by bureaucratic processes. This dictum applies equally to all forms of government, from dictatorships to the most well-intentioned democracies. Almost invariably, these man-made rules become corrupted by alternative agendas, and develop a slant to sustaining those who make the rules, rather than seeking an ultimate benefit for society. One could be forgiven for making the cynical observation that there might always be in place a clearly defined set-of-rules solely to accommodate the whims of the ruling classes. That said, many of these rules are quite essential for the smooth running of society.

There are examples from history where the rules of man run into conflict with the natural laws. The Mayans and Incas developed extremely rich cultures, even though their civilizations were relatively short-lived. Study of cause-and-effect was in its infancy, and some interpretations suffered high error-rates. Natural laws usually provide an opportunity for graceful retirement from the population for the old and weak. Much has been written on the practice of offering as sacrifice the young and productive to placate the gods, and this would appear to run counter to the long-term-benefit of a civilization. Perhaps this misguided attempt at formulating rules played some part in the ultimate decline of these ancient civilizations.

Societies are formed around common beliefs, desires, and objectives. These societies gain strength from rules that are both consistent and cohesive. If rules are to be accepted and obeyed without undue enforcement, there must also be a clear benefit to society as a whole. As societies change and grow so do the associated rules. There appears to be a strong correlation between the number of transient or man-made rules, the size of a population, and the ability of that population to communicate. The more we communicate, the more we generate rules. All man-made rules that govern society are experimental, and subject to change over time.

Our species on planet Earth is pushing the bounds on every front as to what the laws of Nature will allow. Any belief that humans now have total control of all factors governing our civilization, is really misguided and incredibly arrogant. Take the natural controls on population as an example. In nature, virtually all animal populations are regulated by the available resources and prevailing conditions. Thanks to Industrialization, Technology, and the sheer weight of numbers, mankind has reversed the situation. Humans are now consuming resources at an unsustainable level, rather than treating resources as a limitation on population growth. There are many natural forces that impose restrictions on population. The decision on where humans could live, and how many could populate a given space, seemed to be one of free-choice. There is a general formula or rule that seems to apply to all life at a high level, both for plant and animal species. Natural mechanisms kick in when population trends are not in the best interests of the planet as a whole. Again, with reference to the Gaia Hypothesis, we are now bending those rules to the limit.

Despite our best efforts, it is very unlikely that these controls can be manipulated to fully satisfy the human inhabitants of Earth. Before we had any influence at all, we were at the mercy of plague and famine. These were large-scale demonstrations of the natural forces designed to limit population growth. Like beavers, we have the ability to terraform our environment and alter the delicate balance to the detriment of other lifeforms. It would seem we have the power of life or death over all life on Earth, except our own. It's quite possible the dominant species in the dinosaur-age developed a similar sense of superiority and arrogance at a time when they ruled the world and their survival seemed assured. Given the limited resources of our planet, we must look very carefully at the management of these resources. In ancient societies, our ancestors tended to live comfortably within the bounds of sustainability.

Many cultures saw a serious purpose in making things better for following generations. Viewed with the benefit of what we now know, some attempts to improve ancient civilizations may seem misguided. But the intent was there, nonetheless. There was no hint that mankind was actually quite insignificant on a Cosmic-scale. However, there is one thing that human civilizations of old agreed upon – they were being driven by external forces, the complexity of which was beyond comprehension. Invariably, as noted earlier, the gods were given all responsibility for the unexplained and the inexplicable.

Looking back a few centuries, there was a common belief that not much existed outside the view from the naked-eye. The stars in the sky were just specks on a canvas; they weren't seen as replications of our own Solar System or Galaxy. Because the conception was relatively self-contained, there was an innate respect for life and the delicate balance of things that control it. Perhaps as a result of a dramatically expanded understanding of the Cosmos over the last few decades, it may seem that our actions have no consequences or significance in the bigger picture. While that's absolutely true, we should not lose sight of the fact that the whole purpose to life is to make things better for the people who are living it, or have the potential to live it.

The above discussion relates to man-made rules. However, these rules pale into insignificance when compared with the far greater number of rules over which we have no control – the rules that govern our existence, the Rules-of-Nature. In defining transient rules, we have a reasonable degree of flexibility. We have no influence whatsoever on the rules-of-nature. In this space, we are mere observers with quite limited degrees of freedom. Although we have no influence on the rules themselves, we can seemingly affect the outcome from the application of some of those rules by altering the situations to which the rules could apply.

Rules-of-nature are not experimental – they have been put in place to take account of all possible future contingencies. They are immutable. And that is an extremely important distinction between man-made rules and the rules-of-nature. From our vantage point in the cosmos, we can view the historical record back almost to the time when our Universe began. On examination of the record over that time, there is absolutely no evidence to show that any of the laws-of-physics have changed a single skerrick. Our model to explain observations of the big picture relies heavily on extrapolating physics from our everyday experience. One thing scientific investigation has established so far is that there is an incredibly complex and mathematically beautiful interconnection between all fundamental components of the physical world. The ultimate goal for science is to find a unifying theory that ties together seemingly unrelated elements. There are no guarantees that such a theory exists at all. We are driven in this direction by the success in the past at finding unexpected relationships between seemingly unrelated physical quantities.

Perhaps there is no better illustration of this than the ubiquitous equation $E = mc^2$. A couple of centuries ago, who could possibly have imagined that energy might be a function of mass and the speed-of-light? There are many more examples, but their examination lies outside the scope of this book. What we can take from knowing that these relationships exist is that there is an extremely delicate balance required in maintaining precisely the environment of our existence. Imagine for a moment that the rules of nature were not immutable, and might still be evolving. Because of the interconnectivity involved, it is not possible to fiddle with one of the rules in isolation without upsetting the equilibrium which is an essential part of everything we know. The empirical evidence available at this time, leads to the realization that the rules we are observing, the rules that govern every aspect of our existence, have not changed since our Universe began. The inescapable conclusion is that the rules that existed at the start of our Universe, are the same rules we see in play today, and there is no evidence that they might change before its end.

We can look back to contemplate a time just prior to the Big Bang, and assume that the Rules simply sprang from some chance vibration in a void. Eventually they might have evolved into the complex set-of-rules that drive absolutely everything. If this was actually an evolutionary process, it would be reasonable to expect that there might be evidence of changes in some of these rules during the past 13.8 billion years. No such evidence has been discovered, even though that period of observation is quite substantial when considering our Universe as the frame of reference. The other thing to be taken into account is the intimate connection between all the constants and rules in the physical world. Any slight change in any one of these is likely to upset the incredibly delicate equilibrium that exists in basically everything – they are so beautifully intertwined and locked into each other that it's simply not possible to change one, even a tiny bit, without dramatic repercussions. What makes these immutable rules so astounding, is that they have remained in place, unchanged since our time began. There has been no necessity or opportunity to vary these rules due to unforeseen consequences in any of the untold numbers of planets that might sustain life throughout the Universe.

The only rules we can change are the rules we make. The suggestion here will be insulting to many – what we might think of as 'free will' can have no more consequence than any particular random result from any of the rules controlling us. On the other hand, we are not living out a predetermined destiny. The path we follow is fairly rigidly contained within the tolerance allowed by outcomes from the immutable rules, but there is comfort to be had from the illusion that we have more say in our destiny than we actually do. As an example, suppose an individual has decided to live to celebrate his 200th birthday. With the best will in the world, it is quite unlikely that this outcome will be achieved.

Humans, as with every other life-form, have a predetermined upper limit on lifespan that remains relatively constant throughout the period that our species exists. Many factors prevent any particular individual reaching the maximum possible age. Life-expectancy for humans has improved dramatically over the last couple of centuries as health and environmental issues are understood and addressed. Genetic inheritance plays a large part in determining whether or not a human can survive to the limit allowed by the blueprint for our species. Although we can affect the outcome to a certain degree by lifestyle choices and medical intervention, the age to which we live without artificial assistance is by no means open-ended.

Many of the rules that seem to apply in our everyday world could possibly be far more complex than first-impressions might suggest. One interpretation is that these immutable rules might appear to vary depending on the scale of the situation to which they apply. Perhaps a more accurate way to view this is that the rules are not changed at all, but rather we failed in our initial observation to understand the full extent of the rule. When can we be sure that the rules have been particularized to the fullest extent into fundamental components? Answer – "We can't". And a similar frustration awaits particle-physicists as they delve ever deeper into the subatomic world.

A good example of how the same rule might have different consequences in micro and macro situations, relates to the most common force that impacts us all – Gravity. Popular science teaches about the phenomenon whereby objects with mass attract each other. Simple, yes? Well maybe not quite so simple. Dramatic experiments in recent times are just starting to reveal the complexities related to mass itself (ref: **The Higgs Boson**). So the Rule or Law for Gravity is as the textbooks say when considering objects visible to the naked-eye. However, it is not valid to assume that the observations made within the limitations of our experience will automatically extrapolate to the subatomic or cosmic scale. We cannot be sure the Laws of Gravity scale up to cosmic levels without additional, as yet undiscovered, rules kicking in. Gravity is an extremely weak force when compared with other forces in Nature, and decays rapidly over distance. The jury is undecided on whether or not gravitational forces as we currently understand them, are sufficiently strong to draw together tiny particles of dust and gas from the far reaches of space, and create new galaxies. Undoubtedly there are many unknown rules awaiting discovery in this area.

As another example of a rule that appears to produce different results as the scale changes, consider the electrostatic repulsion between similarly charged objects in our everyday frame-of-reference. In physics, this phenomenon is known as Coulomb Force. If this electrostatic repulsion was the dominant force between charged particles at the subatomic level, protons would be unable to collaborate to form stable nuclei. One of the four fundamental forces of nature is known as Strong Nuclear Force. Until humans had the tools to provide an intimate view of internal atomic structures, this force remained hidden. It operates at exceedingly minute distances, and produces no discernible effect in the visible world. Although Coulomb Forces may be present at the atomic level, the effect they produce will be undetectable in any competition with Strong Nuclear Forces.

Rules can produce quite different outcomes as conditions change. Take as an example the rules that govern elemental Carbon. At normal temperatures and pressures, pure Carbon exists in a state that we might find in burnt forests, or pencils. Under intense heat and pressure, diamonds are created from the same material. And we are just discovering the incredible properties of another form of Carbon – Graphene (ref: **Graphene**). This Graphene is worthy of special mention for plaudits to the power of Science and Engineering. It does not occur naturally, and requires a very controlled manufacturing process. However, it is important to note that, even though the process does not occur in Nature, the rules that allowed for it to exist were already defined long ago. This was a 'discovery' by mankind, not an 'invention'.

A common theme throughout this book emphasizes that there are relatively few rules in nature we have identified that can be categorized with certainty as absolute. Like most common compounds, Water seems to follow the rule that it will expand as the temperature rises. The rule for water involving temperature and expansion, has qualifications attached. At both the high and low end of the temperature-scale, water changes state and thus needs a different set of rules to define its behaviour. As the temperature drops and water starts to turn to ice, it expands, in contravention of the previous rule that might have been applicable more broadly. This illustrates nicely that while some rules might appear to be flexible, that flexibility could in fact be the result of macro-rules being composed of very specific 'sub-rules' which have a more restricted application.

To illustrate the delicate interplay and interconnection of rules that determine the perfect conditions for life as we know it, we need look no further than the chemical properties of H_2O, and their role in what we know as "weather". Evaporation and precipitation result from properties of the water molecule, and ultimately control the distillation and distribution of this molecule that is so essential to life. It could be argued that the distillation inherent in the evaporation process is not an essential to life, but it is certainly essential to human life. Sea creatures have adapted to both freshwater and saltwater environments, and presumably humans would have evolved to fit with whatever the prevailing environment provided.

Many of the rules-of-nature are fuzzy rules, meaning they appear to have non-exact outcomes. Immutable rules are not necessarily exact rules – there is no guarantee that the same circumstances will result in the same outcome. Every rule in Nature has a probability of returning a certain result. Some results are quite constrained and predictable; others produce outcomes within a wide tolerance.

Quite different results can be obtained from the application of the same rules in quite similar situations. These variations can sometimes be attributed to a change of environment, or varying initial conditions and properties. Discovering what these influences might be, is the never-ending task for human science. Look to Chemistry for analogies on how these rules, properties, and environmental parameters might interact to produce a specific observable outcome. We might note the result from a given experiment, and hypothesize that it may constitute a general rule that could apply in other situations. If empirical conformation supports the idea that the rule is reasonably definitive and could have wider implications, then that rule is offered as a permanent member of the science of the day. The ultimate success of such a procedure would come if a rule attains status to be recognized as one of the fundamental laws-of-physics.

The conventional way of conducting experiments in Chemistry is to combine various elements or compounds and search for patterns in the results. From those, it may be possible to define properties that can be associated with these elements or compounds. Not only can we vary the elements input to an experiment, we can change the environment substantially with variations in temperature and pressure. Thus, any rule we wish to formulate from the results, must specify the conditions for that rule to have a replicable function.

All physical objects in our Universe are composed of atoms, and hence these objects should be expected to at least obey the rules associated with their atomic components. As atoms combine to form larger and more complex entities, additional rules are necessary. For example, the rules identified as dictating behaviour on an atomic scale would not be sufficient to explain the physics related to black-holes. Nor would they be able to provide any meaningful contribution to our understanding of carbon-based life forms. At what point do the rules governing an object's behaviour extend beyond just an amalgamation of rules related to atoms?

And this poses the question as to where the rules might actually be stored. Is it possible that each atom possesses information about every conceivable outcome involving atoms? This concept is unlikely in the extreme, so we must seek an alternative explanation. One possible conclusion is that the rules must pre-exist outside any object, rather than being self-contained properties and functions belonging to that object. We will examine this concept in a later chapter.

As mentioned previously, mankind has the power to change the planet, for better or for worse. The issue of Global Warming has now shifted from debate, to one of conclusion. It would seem we are largely responsible for the out-of-cycle warming our planet is facing. As we have done in the past, we might look to technology to save the day. If we divert a large portion of the planet's resources to driving refrigeration-units, we might prolong human comfort for some time. Eventually, there will be no escaping the damage that such a temporary misuse of resources will have on the planet as a whole.

We would all like to believe that our species is in some way special. If we look at the timeline for our Solar System, and the eventual burn-out of our star, there will actually be many opportunities for human civilization to be obliterated, and completely replaced by a new dominant species. One can only guess whether these new civilizations start completely from scratch, or whether we leave behind something of interest in the fossil record that might be a good launching pad for Humans-II. The evolution of life and planet Earth will be following exactly the same rules as have applied in the past. However, the countless variables that play a part in determining the future will mean that the history of life on planet Earth may not necessarily be a good indicator of what lies ahead.

RULES FOR ETERNITY

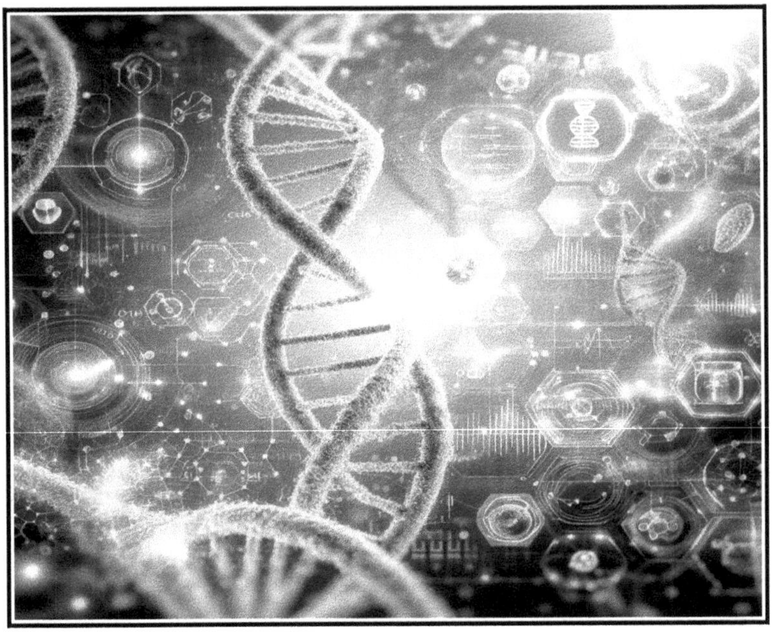

2. BIOLOGY – THE RULES FOR LIFE

Rules that govern the physical world are wondrous, complex, and forever beyond our complete comprehension. Add to that the mysterious property we recognize as life-force, and the level of complexity increases many-fold. From the smallest observable subatomic particles, to the outer reaches of the Cosmos, every aspect is controlled by a predetermined, hierarchical set of rules. At the lower level, subatomic particles follow their own set of rules which allow them to form atoms of various kinds. Another set of rules dictates how atoms might combine to create molecules, possibly of great complexity. One of the most intriguing structures to be built from these molecules is DNA, an essential component of all living things. And again, DNA chains or chromosomes have a unique set of rules that govern their behaviour. Chromosomes exist within cells, and embryonic cells can continually divide to create the many and various forms of life we recognize.

Every living cell must be driven by a common set of rules. It makes no difference if the cell is destined to become a dog or a dinosaur, it must still obey the same rules. So, if all cells behave according to the same set of rules, what determines dog or dinosaur? The obvious answer must be that a cell also contains its own unique information, which in turn influences outcomes from the generic rules. A truly remarkable feature of the life-cycle is the mechanism of reproduction that allows accumulated relevant life experiences to be passed to future generations. This information is stored in DNA sequences. Here we are trying to make the important distinction between Rules that apply equally to all entities of the same type, and Properties or characteristics that only influence outcomes for specific entities.

Every living organism is the product of both the rules that govern life, and many of the properties inherited from the single cell that continually divided to create the organism. Any condition or state that makes a particular entity significantly different from its peers, is considered a Property of that entity. It is not always easy to distinguish between what might be considered a Rule, and what constitutes a Property. Information programmed into DNA chains relates to both individual characteristics and species specifics. As an example, consider a frog born with some psychological abnormality that sets him apart from his peers. DNA instructions have determined that he was going to grow to be a frog. Embedded elsewhere in those instructions could be some indication of psychological problems occurring somewhere along the cell-division process.

Over the last 50 years, biologists have cleverly and convincingly shown how characteristics of various species adapt to their circumstances, and pass on the information to benefit future generations. The way DNA replicates and stores information is central to the whole mechanism of evolution. However, there may be additional ways of passing information from parent to progeny rather than direct inheritance or teaching by example. Perhaps we have assumed there can be no information transfer from mother to child at or before birth, just because we have no explanation of how such a process might work. Instinct is something that many animals, including humans, appear to possess at birth. Is instinct a pattern of behaviour that belongs to a species as a whole, or does it result from collective wisdom passed down from previous generations? It is not clear we currently know the answer to that.

To be consistent with the premise that all things ultimately result from immutable rules, we need to identify which rules apply equally to all forms of life. One of the most obvious influences on life, is death. Even though end-of-life is often a sad experience in the human context, death is an essential part of life. The approximate end-game for each individual of a species, is just as carefully orchestrated as the start of life. One of the most challenging questions for biological science, is why we age, and what are the mechanisms causing us to wear out. Recent biological advances have given some insight into the way DNA chains are continually shortened at each replication (ref: **Telomeres & Immortality**). There is some randomness associated with this shortening process, and obviously some lifespans are longer than others. There are environmental factors that influence this, but there seems to be an underlying constant built into the genes of every living organism that ultimately dictates a certain natural life-expectancy. It is unlikely we will be able to manipulate the rule that drives this shortening process. However, what we do have is the ability to medically interfere with the natural telomere mechanism, and thereby potentially influence life-expectancy.

So why is programmed mortality important in the survival of the species? The obvious answer is that there must be a sustainable balance between the creation of new life, and the phased extinction at the end of each life. In addition, there could also be a more subtle explanation for the phenomena. A requirement for Evolution, with its dependence on Natural Selection, is that there must be a selection available. This selection results from chance mutations that propagate through a species. Mutation also occurs as cells replicate within an organism throughout its lifetime. This mutation is a double-edged sword, and often produces very undesirable outcomes. As a direct consequence of mutation, various forms of cancer can develop as an individual ages. With programmed mortality, and less likelihood of participating in procreation late in life, there is a safety mechanism that reduces the risk of genetic flaws flowing through to the next generation.

There is another mechanism capable of keeping the population of specific species within the bounds of sustainability – Regeneration. This process is rarely found in either plants or animals. Studies of a particular species of jellyfish, Turritopsis Dohrnii, have demonstrated that they have the most amazing ability to return from adulthood to an infant state, under the right conditions (ref: **Turritopsis Dohrnii**). However, if regeneration was common throughout all lifeforms, it would seriously impede the evolution process by limiting the cross-pollination of gametes. We might interpret the curious behaviour for this particular type of jellyfish as an aberration that is allowed within the rules that govern all life.

The notion of immutable rules mentioned in the previous chapter, seems to have more relevance to the rules of physics, than to the rules of life. However, this does beg the question as to whether the rules regarding life might also be immutable. The possible outcomes from applying the rules for life seem endless. When we apply the rules of physics, it appears as though the results range from a clearly defined set of atomic structures, through to relatively predictable behaviour at a cosmic level. What is sometimes obscured in our macro view of the world, is that even the laws-of-physics can produce many different outcomes from the application of a single rule. Every rule has an associated probability that a certain outcome will result from certain input conditions. When we drill down below the atomic level into the strange realms of quantum mechanics, there is no requirement that any particular particle has to be either black or white – it could in fact be black AND white at the same time.

We can identify the chemicals and molecules that are essential to life. What we have more difficulty with is deciding where and when the rules associated with life come into effect. At some point after the creation of our Universe, atoms combined to create complex molecules containing the blueprint for life – DNA. The cells containing DNA had acquired a momentous property in the evolutionary scheme. We know approximately when this happened, about 3.5 billion years ago, and are coming closer to understanding how. As regards rules, atomic structures that had been governed purely by the laws-of-physics, now in some mysterious way obey an additional set of rules that belong to the branch of science we term "Biology".

Delving deeper for an understanding about which rules might be associated with which entities, we are faced with the same dilemma as when deciding at what point a collection of atoms might eventually become living matter. Stem-cells, almost spontaneously, begin to follow rules related to the role required of them in a specific environment. They seem to develop and grow entirely in response to their situation. It's interesting to examine the mechanisms involved where they follow both the rules that apply to all stem-cells, and also obey an independent set of rules specifying a group dynamic. Indications are that new cells receive instructions in the form of chemical signals from surrounding cells, encouraging them to perform in a role that will be of benefit to the group as a whole (ref: **Stem-Cell Niche**).

At the moment our Universe was born, most scientists agree that visible matter consisted mainly of simple atoms like hydrogen and helium. Complex atomic structures were formed at a later time during the lifecycle of stars and galaxies. It is extremely unlikely that any complex molecules existed at the time of the Big Bang, and it must be assumed the material defining life has been incubated elsewhere at a later time. Even though the rules pertaining to life existed at the time of the Big Bang, the environment conducive to life was a long way off.

To try to understand this idea of rules pre-existing any circumstances where they might be applied, let's consider a hypothetical analogy. Imagine millions of Lego blocks were poured into a giant mixer. Left for long enough, the plasma of bricks could fall into place and form Lego-Land. It is easy to dismiss this outcome as absurd. And yet, we readily accept the idea that planet Earth and all human civilization has resulted from exploding stars spewing particles into the galactic mixer. When the first Lego bricks were created, the designer could not have envisaged all possibilities resulting from their use. Similarly, the guys who built the giant mixer in our analogy, could not have imagined the device being used to organize Lego bricks into meaningful structures. So if Lego-Land was the result of this experiment, there must have been extra input besides just the bricks and mixer. This extra input must be quite independent of the building blocks or the environment.

Continuing with this analogy, let's contemplate what makes the Lego-Land result so unbelievable. Comparing the experiment with the functioning of our Universe is grossly unfair for three basic reasons. Atomic particles are closely integrated with the laws-of-physics, almost as if they were originally part of the same design specification. Anything involving Lego bricks must rely on a tiny subset from the laws-of-physics, without benefit of co-working with these laws. Secondly, the experiment is seriously disadvantaged by the fact that it is only being run once, and thus there is no opportunity to modify the original design to cope with unanticipated situations. And thirdly, formation of Lego-Land requires something not accessible from an environment composed solely of plastic blocks – Intelligent Life. In engineering terms, there is something magnificent about any design that stands the test of time, and behaves predictably in every possible future scenario. From a human perspective, this is impossible to achieve. Because the laws-of-physics and the atomic building blocks have remained unchanged for 14 billion years, one could be led to the conclusion that our Universe is not the first where these rules have been trialled. Alternatively, maybe the rules existed unchanged for eternity.

Returning to reality, it is hard to build a model for our existence that does not require input from some unseen and unknown source. Our understanding of atomic structures that are fundamental to everything in our physical world, is expanding at an incredible pace. It must be remembered that humans can never extend their role beyond that of observers. We can conduct some pretty amazing experiments within the existing laws-of-physics and Nature, but with the best will in the world, we can never create a giant mixer that 'knows' how to construct Lego-Land. Admittedly, some exceedingly convoluted engineering might do the trick, but that would still require physical implementation. In keeping with the theme of this book, we have categorized the mysterious inputs as the Rules for Eternity.

Interplay between the rules for life and the broader set of immutable rules, creates some interesting possibilities as to what form life might take in extreme alien environments. What if the rules-for-life themselves are far more flexible than we currently understand them to be, and any limitations are imposed solely by mechanical constraints and the laws of physical science? Taking examples from our home planet, it does appear that over a very broad range of environmental parameters, life will attempt to establish a foothold. Because we are surrounded by carbon-based life on Earth, is it valid to assume that this must be the only formula for life? Possibly not. Examining the constraints of physics on the life we are familiar with, usually provides a logical explanation for the way a particular life-form is implemented. For giant trees, the tallest living things, gravitational forces dictate that their centre-of-gravity must be situated close to the centreline. Any serious deviation that creates an imbalance will cause the tree to topple.

As another example of how the laws-of-physics greatly influence what form life can take, consider how huge some species of dinosaur had become during their reign. Limitations were imposed on size, not just by how much food there was to eat, but also how much of it they could eat. If they were purely eating machines, surrounded by food, survival would not be driven by how fast they could travel – a small coefficient-of-drag did not play significantly in their design. These massive objects were seriously hampered by inertia. Perhaps they might have been able to grow even larger but were prevented from doing so purely on the grounds of logistics. Their huge bulk could have made them so lethargic that they became increasingly reluctant to make the effort to drag themselves far enough to find a mate – a fatal mistake in the plan for survival of any species. On a different planet, with quite different gravity, dinosaurs could have evolved to be similar shapes, but radically different dimensions.

Of particular interest to us humans is consideration of what might be limiting factors on the survival of our own species. All animals, with the notable exception of man, surrender gracefully to the dictates of nature that have been designed to ultimately benefit the whole planet and all forms of life. Outcomes might seem harsh on an individual basis, and entire species can become extinct to be replaced by new players. However, that is the broader cycle of life that we have been born to, and must accept. Humans have judged themselves to be positioned at the top of the intellectual tree on planet Earth. Note that other species might assess human intelligence somewhat differently, and could have made a collective decision not to openly challenge the claim. With unmatched ability to influence the planet on a large scale, it is understandable that humans might be tempted to believe they can operate outside the laws of nature. As mentioned previously, this is well demonstrated by the human population explosion over the past century. Population density, distribution, and diversity have grown beyond the point where outcomes could be controlled by a handful of powerful leaders. There is now a degree of inevitability associated with the fate of humankind, and it certainly incorporates many undesirable components. One rather pessimistic interpretation might be that human civilization has run its course, and now faces extinction according to the same rules that apply to all species.

Thanks to careful observation over many generations, guidelines for the good health of both individuals and planet Earth have been established. In recent times, science and medicine greatly sped up the trial-and-error learning process. One recurring theme relates to what we eat. Looking back as best we can over thousands of years, it seems dietary patterns of old might have produced more healthy individuals than is frequently the case in some pampered Western societies of today. There should be no surprise here, and the explanation relates to the slow rate of evolution of the human animal. Back when our cave-dwelling ancestors were evolving to suit the food available to them, the basic bio-mechanisms were established. Because changes in lifestyle were extremely slow, there was time for evolution to keep pace with mild environmental changes. In modern times, diets are changing rapidly and dramatically, and not always for the best. The biomechanisms in place today have not really had a chance to adjust, and research often indicates the benefits of returning to dietary habits of the past. One of the most serious threats to human civilization is the fact that the environment is changing more rapidly than we can adjust via evolution.

It is almost impossible to distinguish between effects from the fundamental rules of life, and those that result from environmental conditions or influences from other life-forms. Various species might compete in the same territory for the same resources; others might be complementary and provide support and new opportunities. There is one function that does appear common to all forms of life – Optimization. Natural selection that plays a role in evolution is one of the more obvious forms of optimization. Once the basic requirements for life are met, as they are here on Earth, there seems to be no restriction on where and when a new species can be created. The form it takes is designed to fit exactly within the resources available. That surely must be the perfect example of optimization at its best. Virtually every property of plants and animals is subject to optimization.

As an example, consider what factors influence the average height of humans. Over the last century, average height has increased a little in response to improved nutrition. There were noticeable dips in height related to both world wars, possibly resulting from short-term poorer dietary conditions. Studies have shown that the average height of humans today is slightly less than it was around 150,000 years ago (ref: *Human Evolution*). The rate of increase today is slowing, suggesting there might be an inbuilt limit for the optimal height of man, and that limit was set when Homo sapiens first appeared. It is logical that early humans were optimized as hunters, and being tall meant they were able to run fast. The idea that there might be age-old inbuilt constraints on humans has quite interesting implications.

Enhanced capabilities for collecting and analysing huge volumes of data have helped identify the subtle chance-influenced adjustments brought about in the process of optimization. As we are discovering with ever-increasing clarity, anything we can observe, measure, and interpret as a fundamental rule of nature is most likely an encapsulation of various other rules. Consider the rules that human science has ascribed to the whole spectrum of molecular chemical processes. They may seem to be relatively independent, deserving of classification as fundamental rules. However, all chemical processes and interactions are ultimately dependent upon the basic properties of atoms. Once again we are confronted with the mind-blowing realization that atoms were originally contrived in such a way as to enable the beauty, complexity, and vastness of existence.

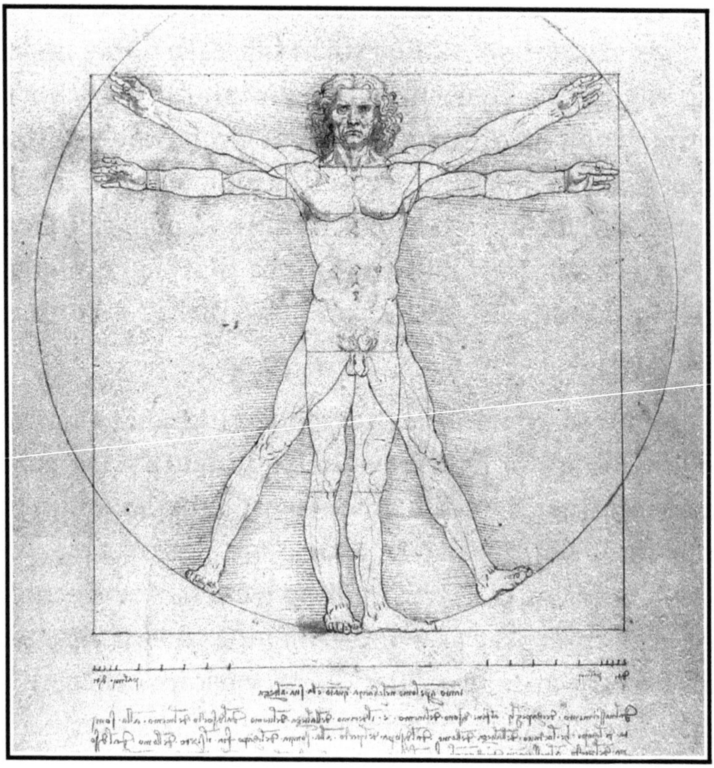

3. THE HUMAN FACTORS

"For Planet Earth to be in sustainable equilibrium, the average human must be 'resource neutral'. Most botanical life already obeys that rule."

The previous chapter made reference to the fact that all living things have an average life-span associated with their species. For humans, there has been an overall steady increase in life-expectancy through the ages. This has come about because of improving conditions, and in more recent times, medical intervention. As our life-expectancy increases, new threats are revealed that did not have enough time to influence the lives of our shorter-lived ancestors. In the past century alone, there have been several challenges to health that have been addressed to a large degree. Firstly, disease and viral infections were tackled by major advances in medical science. Next came the threat posed by cardiovascular issues in the ageing human. Education and lifestyle changes had a significant impact, as well as medical intervention in the more serious cases. Today, the biggest limiting factor on lifespan appears to be Cancer in its various forms. The unanswered question is just how far can we push the average life-span for humans, given that the original optimized design for them was drafted long ago.

Referring to human-life as a separate entity is a huge oversimplification. Any individual human is actually teaming with microorganisms that cooperate in a symbiotic relationship with the basic human components. Inside the human body, there is a never-ending battle between the positive elements and the negative. Eventually, the negative elements overwhelm the various defence mechanisms, and life comes to an end. This same battle plays out in all realms of life. So when we refer to the original optimized design for humans, it is important to incorporate the fact that the microorganisms that inhabit humans today, might have evolved to have very different characteristics and functions as when they first became linked with humans.

Application of the rules pertaining to life produces an unimaginably large number of diverse possibilities. However, the possibilities are not infinite. As with all else in existence, the resultant entities still must comply strictly with the constraints imposed by the laws-of-physics and chemistry. So let's examine the broader parameters that drive life-as-we-know-it. One of the fundamental controls for life is temperature-range. Each particular cellular structure or organism has a certain temperature-range for survival. As humans, we are doomed if we stray beyond the bounds of our temperature survival-band – at the low end of the scale we freeze, the other, we fry. Somewhere in between is a comfortable balance where we thrive and multiply. Most researchers believe that around $150^{\circ}C$ is the theoretical cut-off point for life. At that temperature proteins fall apart, and biomedical reactions cannot occur – a quirk of the biochemistry that life on Earth, as far as we know, abides by.

To a large degree, Earth's atmosphere controls what forms of life can exist on the surface of the planet. It provides a mechanism to supply the essential elements required to sustain life. Chemical composition of the atmosphere is key, but there are other atmospheric properties that have an influence on life – primarily Pressure and Temperature. External pressure from the surrounding environment determines many of the internal characteristics of animals. Some of the more exotic lifeforms on planet Earth exist deep in the ocean where there are vents through the Earth's crust (ref: **Thermal-Vent Creatures**). As well as the extremes of temperature and pressure, these hardy creatures coped with a toxic chemical environment that would have excluded previously known forms of life. The comfortable existence we enjoyed on the surface of the planet, encouraged the belief that such conditions were essential for all life. These creatures completely rewrote the book on what we imagined would be limitations for life.

One of the more obscure and sometimes overlooked aspects controlling life is background radiation. If that exceeds a certain critical level, we would cease to exist in any form that we might recognize. There seems to be an influence on life from radiation, both particle and electromagnetic radiation. At this time, the extent of that influence has not been clearly established. One of the more radical suggestions is that perhaps the radiation continually bombarding Earth plays a part in our evolution by causing the mutation of genes. In addition, there are the undesirable effects of home-grown radiation, and these have been well documented.

Up until this point, we have been developing the theory that everything in existence is driven by a predetermined set-of-rules. This concept becomes most challenging when we attempt to examine the origins of complex lifeforms like humans. Humans are intimately woven into the intricate fabric of life, and hence have a special interest in trying to understand the purpose behind the rules, and where these rules might ultimately lead. The rules defining life as-we-know-it affect the full gamut from single-cell beginnings, all the way through to the recognizable human form. Because humans enjoy maximum flexibility and freedom to play in three of the common four-dimensions, they possibly might require many more rules to control their existence. The difficulty here is to determine if there are specific rules for human life which differ from the generic rules that apply to all forms of animal life?

Posing the question another way, "Is there a specific blueprint for humans; for human life?". In attempting to find an answer, let's imagine a human is composed of basically two parts; the physical-human and the psychological-human. Much of the physical-human can be explained by the generic rules governing evolution for all life. The psychological human is more difficult to wrap in a precise definition, and is related to brain activity and the resulting behaviour. Human behaviour can sometimes be identified in animals, and thus it is not always easy to determine which traits are uniquely human. Emotion is something often seen as a characteristic belonging uniquely to humans. Darwin first raised the issue of animal emotions over a century ago, and it has been a hot topic for debate ever since (ref: **Emotion in Animals**). As medical science discovers ever more detailed information about which parts of the brain are responsible for what function, we may be nearing the point of determining whether animals have emotions or not. If there is evidence that human characteristics, like emotion, can also be found spread among various animal species, it would become increasingly hard to argue that mankind is special and subject to its own set of rules.

In spite of spectacular advances in the biological sciences over the last few decades, many more questions remain than have been answered. As yet unexplained is the formula common to all animal lifeforms – "What rules govern the choice of partners to ensure the survival of the species?". One of the fundamental rules for life is that different species can originate from a common ancestor, and yet those species remain in a separate reproductive stream with little crosstalk between species. The consequences of not having such a simple rule for separation would be disastrous. There is another more subtle rule that dictates partners for procreation. To many, the idea of discouraging sex between siblings or close relatives might seem like a man-made rule that has become institutionalized, rather than a more fundamental rule supporting survival of the species. Over time, inbreeding prevents the diffusion of genes that is so essential to healthy evolution. In turn, this can lead to illness or disabilities that discourage future participation in continuing the bloodline. Somewhere these rules have already been defined and we are just playing along. Thank heavens the rules have been so carefully crafted. Otherwise, chaos would reign.

There is a simple observation when comparing plant and animal life on our planet. Plants have a symmetrical randomness, and their construction results mostly from the need to obey the laws-of-physics. Take as an example the way that the branches of tall trees are distributed around a vertical axis. Although branches on one side of a tree would rarely mirror branches on the other side, there is nearly always a balance in the way the branches are distributed around the vertical axis. If too many branches formed on one side of the tree, resulting forces on the root system could overwhelm its anchoring ability, and remove the tree from future participation in evolution of its species.

Animals on the other hand, have a very distinguishable external bilateral symmetry in their design. Bilateral symmetry makes good sense when considering how important balance is to animals with the freedom of movement. Any lopsidedness in an early mutation would have evolved out of a species, purely on the grounds of survival. The real problem arises when we look at the internal design of animals. For organs that consist of two parts or are duplicated, like lungs for example, good design suggests locating these organs around the lateral centre-line. Many organs in animals are one-offs, like the heart for instance. There is no way to fit all these on a centre-line to maintain good symmetry, and therefore there must have been very careful planning on how to fit all these elements within a given structure. Again, taking the heart as an example, this sits exclusively on the left side of the chest in humans, except perhaps for the occasional mutant. And the burning question is, "Why?".

This question about alternative designs is equally valid and mysterious for all animal-life, not just humans. The template for all animal species has a clear, asymmetric layout for most internal organs. Did natural-selection play a part in deciding the internal organ structure for humans? Given their limited knowledge of anatomy, would it be sensible to speculate that some early cave-dwellers might have been more attractive to sexual partners because of the location of their internal organs? In previous situations we've examined, there is usually a reason for each and every rule-of-nature that we can identify. Could the phenomena described here be an example of a completely arbitrary choice that became a blueprint for all life that followed? Life is bursting forth spontaneously and independently right across planet Earth, and there is not always a tie back to any single ancestor for any particular species. To be consistent with this observation, the rules governing life would need to specify preference for a certain format, even though there may be no identifiable advantage.

To further investigate what might be missing from our picture of mankind's evolution, think about possible alternative designs for man that would not have altered the species in any significant way. For instance, we might have had a heart on our right-side instead of our left, or maybe even two hearts. The laws of evolution dictate that the best choices eventually win out over options less conducive to long-term survival of the species. But what happens when there are options that are absolutely equal in terms of benefit to the species? Would it be reasonable to expect that both alternatives survive and develop in parallel? It is hard to imagine why having a heart on one side of the body would be a better option for survival rather than having one on the other. And yet, there seems no evidence that evolution experimented with these alternatives during the earliest manifestation of mankind. This does seem to offer a subtle challenge to the mainstream theory of evolution which focuses on the selection of options which provide a discernible benefit.

In our quest to try to understand how the system of rules might fit together, we now find it could be necessary to accept there are some rules which are arbitrary, that have no additional benefit one way or the other. Back to the logic behind having the animal heart on a particular side. The basic layout of things anatomical has a common theme throughout most of the animal kingdom. We could believe that all life relates back to a single ancestor. If you accept this theory, then there must also have been parallel experiments that were discarded. Normally we should not expect the rules to give preference or bias to any particular animal blueprint, without there being an associated benefit. Could there be some bias written in the rules? The concept that we will develop a little later on is that all rules have purpose. And yet, here is an example where the rules seem to be enforcing one choice in preference to another, rather than giving equal opportunity to multithreaded, equally-viable solutions.

Another school-of-thought has it that the whole of everything resulted from Intelligent Design. While it is not the intention here to give weight to that idea, it is difficult to dismiss that concept entirely. Whatever understanding we might claim about our origin, two things are hard to dispute: everything has been magnificently designed, and designed intelligently. And the point that differentiates between all interpretations... Who or what was responsible? The conclusion we would hope to make obvious at the end of this book, is that nothing was responsible for creation because there was no actual beginning. We have quite an insignificant role in the grand scheme of things. The playground where this incredible exercise takes place has existed forever, and will continue to exist forever, regardless of whether there might be lifeforms participating in and observing the process or not.

Let us explore the rules that might contain blueprints for life. If we believe there is a blueprint for every possible life-form, then that leads to the conclusion that there is a near-infinite number of blueprints. A more plausible concept is to perhaps deal with a smaller number of blueprints that apply when certain preconditions for life exist. If the blueprint for man as he exists today was drawn long ago, it might be possible that man could jump the evolution-queue. That might fit well for Creationist Theory supporters. However, the fossil record on Earth does not support that notion. One interesting analogy is to think about the way Science and Technology have developed. Advances-of-the-day draw heavily on preceding discovery and knowledge. For instance, we could not have had mobile-phones before the invention of the transistor and large-scale integration of electronics. We might map that to the grander scheme, and conclude there is a requirement for certain life levels to exist before specific rules can be applied to facilitate the next level of life.

So far, we have covered situations that are governed by rules. It is equally important to investigate areas where there appear to be no rules. Much of human endeavour is devoted to artistic and cultural development. It is hard to argue that the direction of such development is preordained, rather than a free-will application of chance. As an example, let's look at music as an art-form. What constitutes good music, bad music, or just plain noise, is very subjective. Could there be rules built into the human animal that allow it to distinguish music from noise? In physics, most objects have a natural frequency, and it seems humans may have a similar property. If we analyse marching music – there is a particular rhythm and frequency ideally matched to a fast walking pace of the average human. The suggestion here is that, even though we have complete freedom to create music, there could be some quite obscure rules giving bias to cause certain types of music to be more appealing than others. Widespread rejection of avant-garde styles of art suggests that familiarity is paramount, and that familiarity could result from exposure to things in nature that are in turn controlled by rules.

In the broader context of artistic endeavour, it is hard to imagine any influence from some background rules would be detectable against the influences of chance. Because artistry and culture are driven almost entirely by random creativity, there is an exceedingly remote possibility of there being another William Shakespeare pumping out high-quality literature somewhere on one of the trillions of Earth-like planets.

Let's project way into the future when our Sun is nearing the end of its life, and temperatures on Earth might require completely different characteristics for carbon-based lifeforms. Thankfully there are several billion years before we need to be overly concerned. There would appear to be two limiting factors regarding rising temperature and life as we know it – Water and Atmosphere. Once they have evaporated from our planet, there seems to be very little chance for carbon-based life to survive. Recent data from Mars missions strongly indicates what our future might be when we lose these two essentials for life. If Earth maintains its current orientation and spin, we might expect there to be some temperate zones that are more conducive to life than others. And these zones would change location on a regular basis. Thus, we might expect that animals with the best ability to migrate, would be the most likely to survive. The ocean depths may also provide respite from the searing surface temperatures. Large surface-based creatures would be the first to go. Heat dissipation might become a deciding factor – skinny creatures with large surface-areas in proportion to their volume, have a better chance of staying within the temperature band best suited to Carbon-based lifeforms.

Planet Earth has existed nearly half as long as our Universe. This has provided ample data for scientists to look back and study results from the many and varied experiments with life. It has also created an excellent opportunity to identify factors common to all forms of life. One remarkable conclusion to be drawn from the evidence so far, is that all life appears to be carbon-based. When dealing with such complete unknowns as the possibilities for extraterrestrial life, it is never a good idea to say 'never'. In predicting such things, we are constrained between "very likely" and "very unlikely". It would be a reasonable assumption to expect carbon-based life forms similar to those found on Earth, to exist somewhere out in the vast expanses of space. What we can be less certain about, is the possibility of non-carbon-based life because there is nothing here on Earth that gives any hint that such things are credible. Trying to define what separates life from nonlife is difficult, but one clue is in the way that Darwinian evolution belongs exclusively to living entities. There is no evolution involved with Chemistry or Physics – the processes that existed billions of years ago are identical to those observed today.

No matter how thoroughly we investigate the origins of life, it is impossible to avoid the chicken-and-egg question. Did the template for life arrive in our solar system the same way heavy atoms were delivered – from beyond? That theory has a significant number of subscribers, and recent data from the Rosetta mission indicates at least some organic material may have originated outside the bounds of planet Earth (ref: **The Rosetta Mission**). Alternatively, pre-existing rules defining where life might be allowed to form could have been just lying in wait for the right conditions. In other words, there was no requirement for the chicken to pre-exist the egg. But there is an unmistakeable message here; the RULES permitting creation of both chicken and egg must have existed long before presenting a paradox for philosophers.

There are as many guesses as to what form life might take in other solar systems, as there are science-fiction writers. One proposition that will likely remain unchallenged is that the possibilities are endless. Life on all the planets, in all the galaxies, is driven by the same rules, so the diversity we see on planet Earth is merely a glimpse of what may happen elsewhere as a result of varying conditions and chance.

So putting all these factors together really highlights what a delicate balance there is to create exactly the forms of life that we know about. Unfortunately, this does not offer any clues as to the possibilities for life beyond what we might imagine being the norm. Given that essentially the same circumstances as we have here on Earth might exist in some far-flung corner of the Cosmos, it is interesting to contemplate whether the same lifeforms could have evolved. That's where the rules come into play because they exist absolutely uniformly throughout the space and time of our Universe. It would be logical to argue that the same circumstances would produce similar outcomes, chaos theory notwithstanding. Given the untold billions of opportunities for life in the Universe, it is quite conceivable that this experiment has already been repeated many times. Even given the endless opportunities for life out there, the probability of having exactly the same civilization as we have today, is quite small. There has been a lot of chance variation in our history that could have led to a very different outcome. If the dinosaurs still roamed the Earth, would humans have the same population distribution as we currently do? Would we even have evolved at all, or would our progenitors have been eaten long before evolution had a chance to perform its party-trick?

All things considered, life as we know it is only possible during a minuscule moment between the birth and death of each individual star. It is hard to understand the purpose of expending such a huge amount of effort to create a complex cosmological backdrop solely to support the brief appearance of one particular special species. Developing the theme that the whole purpose to everything within our field of observation is for the benefit of mankind leads one to question the need for such a grandiose scale. If all this was just for man's amusement, it could surely have been done in a more compact way. It would seem completely unnecessary to create totally unbounded opportunities for life in other times and other places, and not take full advantage of those opportunities.

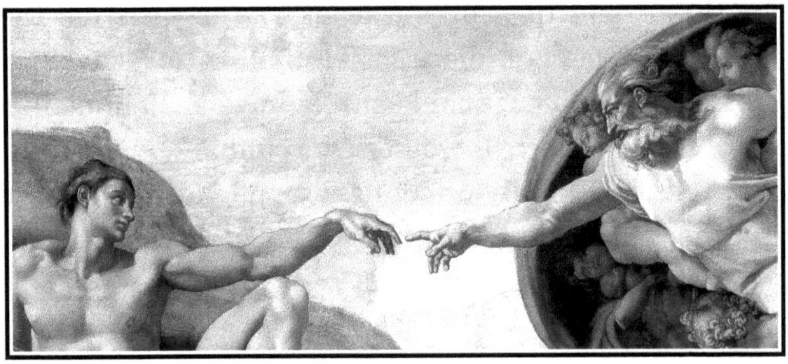

4. WHY WE NEED GODS

Science and religion have one thing in common – both offer the promise to explain things that are currently outside our understanding. The cornerstone for most religions is a God or gods, responsible for all issues of importance. Monotheism categorizes religions that have one God, whereas Polytheism refers to those involving multiple gods. Those gods are seen as creators of the rules that everything in our Universe must obey. There is much comfort to be had in believing one has reached an endpoint that delivers the answer to everything. Scientific investigation and discovery is a never-ending process, and no theory is safe from being obsoleted by future work. On the other hand, religious models rarely face the challenge of revision. Nor is there a requirement that they must. Religious belief is very much culturally based, and there is no contradiction in having many equally valid versions. Science however, has no cultural boundaries and ultimately belongs to all humankind.

Somewhere in every child's life, the time comes when they ask the difficult question – "Where did I come from, mummy?". Mummy has several choices on how to approach this. Perhaps going into all the gory details of childbirth and the tricky gymnastics required

9 months earlier may not be appropriate for a young mind. If mummy had a philosophical bent, she might want to spend the rest of her life trying to answer that question. Even so, there is every possibility she could still be quite unsatisfied by the answer she arrives at. Another approach, and definitely one of the most common, is to look at one of the off-the-shelf religions for a solution. This technique might supply answers sufficient to satisfy childish curiosity, and there is a possibility the solution sits so well that particular children may take that philosophy and grow with it through the remainder of their life. There are many solid psychological reasons why this method may be best, simply because it induces comfort and a sense of well-being in the human animal.

As difficult as the question of where we come from might be, it does not take the prize as the most difficult question of all time. That honour belongs to the completely unanswerable, "Why does anything exist at all?". When investigating components of existence that do not contain life, time-frames of several billion years are involved. An example of this is the formation of the planetary system around our Sun, which is estimated to have taken around five billion years. On the other hand, evolution of particular life-forms occurs within a much tighter time-frame (ref: **History of Earth**). Using the Cosmic Calendar introduced by Carl Sagan, any particular form of life is likely to be only a blip on the radar for a few seconds of the calendar year (ref: **The Cosmic Calendar**). Insulting as it might seem, human civilization is totally irrelevant outside our time and tiny biosphere. It would be quite egotistical to imagine that the grand cosmological design was solely for the purpose of accommodating humankind. Perhaps a more realistic perspective is that all forms of life we know of result from a set of rules that permeate all space and time.

If humankind beats the odds and survives beyond the next couple of centuries, the clock will be ticking very loudly. Many would rail at the idea that every aspect of human achievement could be permanently lost. Engineering, Scientific, Artistic, Cultural – all legacies gone forever. Part of our psyche demands that we have 'purpose'. If we examine all the beautifully complex rules-of-nature that govern our existence, none appear to be without purpose. It would be a logical extension that we humans might also have purpose. What that purpose might be is very dependent upon the context and the time-frame.

Life appears to be quite transient on a cosmic scale. One possible conclusion might be that the rules for inanimate objects could be more refined and appropriate for the long term. If life-forms like humans are destined for extinction after such a short reign, it implies one of two things. Either there are flaws in the design that will lead to self-destruction, or there is no point in continuing with one particular experiment of life in preference to another. To suggest that there could be flaws or unintended consequences related to human design runs counter to the observation that everything in the natural world is indistinguishable from perfection. Although an uncomfortable conclusion, it just might be that humans result from endless chance outcomes, and this idea does fit with the proposition that we're just an experiment devoid of any long-term significance.

So if we are simply pawns in a gigantic experiment, who or what might be responsible for conducting this experiment with a view to analysing the results? Although it is not possible to answer that question in a meaningful way, these thoughts could feed directly into a god-model. This definitely implies a high degree of control and intelligence at the top level. Giving the gods human qualities is a theme quite common to many religions (ref: **Anthropomorphism**). Returning to this idea that man is somehow special, does beg the question as to why he has not been privileged to have much greater access to the big picture. Mankind's involvement in everything is so brief and insignificant, it makes no sense to propose that this entire magnificent scenario was contrived over billions of years, just for the entertainment of a few animated atoms in a near-infinite Cosmos.

If we search long and hard enough, we can usually find an explanation or purpose for every rule or function that applies to the natural world. We are encouraged in the belief that no rule is without purpose. However, it is difficult to reach a similar satisfying conclusion when trying to understand the purpose behind the entire scenario of existence. Identifiable patterns are emerging – whenever intrepid explorers approach the difficult end of the question-spectrum, invariably the god model comes into play. It would seem more important for humans to have AN answer, rather than THE answer. There is widespread adoption of many religions across quite diverse societies and cultures. The evidence to suggest a need for a god or gods, is far stronger than the evidence for the existence of those gods.

To a large degree, the god-model satisfies the need for answers, even though those answers may depend a lot on faith as opposed to reason. Accepting this explanation for the giant mystery-of-existence does provide an opportunity to concentrate on many of the other important aspects of being alive. For some of those humans lucky enough to have a life that they're grateful for, the religious model can provide a focus for gratitude. It's very hard to give thanks into a void – it loses much relevance and meaning. Similarly, it may seem satisfying to sometimes outsource blame for negative aspects of the life experience.

There are other strong parallels to draw between religious models and the structure of the family unit inherent in human society. Most of us are born to be cared for and nurtured. For many religions, there is a strong reliance on the belief that individuals are watched over and guided by some unseen, caring parental entity.

As mentioned previously, gods were invented partly as a mechanism to enforce authority in society, although in all likelihood this would have preceded the notion of the God of the Gaps. Humans, all the way back to their tribal days, have gained strength in numbers, and the tendency to form collectives of the like-minded provided a similar protection to the way animals defend themselves in the wild. If groups or societies developed along religious lines, the notion of their God as protector would help strengthen such groups and reinforce legitimacy. Here was an umbrella figure to provide an extra level of protection from danger, either real or imagined.

Some religions define a much stronger relationship between the gods and Nature, and see the two as synonymous. God is in everything we see and touch. This is termed Pantheism (ref: **Pantheism**). That concept that God and Nature are closely intertwined has great merit, and there are striking similarities. The gods are out-of-reach; entities to be respected and revered, just the same as we should appreciate the magnificence of Nature and everything associated with what we know as 'life'. A fundamental response to the wondrous world we are born to, should involve unwavering respect and reverence for Nature.

Many religions take the approach that humankind is special (ref: **Genesis**). In one sense, man is special because we currently occupy top spot on the food-chain. That ranking will doubtless change over time, as it has in the past. A challenging consideration for those who believe in mankind's permanent status on the 'special list', is to think about the creatures who ruled the planet many millions of years ago. Did they have special status and a connection to the gods? Later in this book we carefully examine the possibility that there are no such things as Singularities. If man is a singularity, a special-case in a special-place, the implication is that he is somehow central in the structure of the Universe, rather than just one example of intergalactic life from near-endless possibilities.

Mankind being special is an ancient idea where the entire Universe revolved around man and his Earth. Today, that model has few followers, and yet we still want to cling to the thought that man is somehow central. Could it be a remnant of the philosophy of old? It is also interesting to look at the template or blueprint for man as suggested by some religions. As discussed in chapter 2, it is difficult to argue against the high probability of countless advanced civilizations existing elsewhere in the Universe. Mankind could be the result of following a generic blueprint, and one might speculate that alien civilizations could be somewhat similar if incubated in similar conditions. Does this mean that out there in another space and time, there are creatures we might recognize as brothers? Accepting for the moment that life does exist in many and various forms throughout our Universe, how similar to humans must these lifeforms be in order to achieve the same special status attributed to our species?

In the same way that we analyse how our physical world can be driven by rules, we need to also question what might drive the less-physical aspects of our existence. Patterns of collective behaviour certainly play a part in the way any civilization ultimately performs and survives. Are there rules written to guide this behaviour? If we had understood these rules long ago, could we have predicted the future as we live it today?

As we gain ever-increasing knowledge about our place in things cosmological, it becomes clear that 'forever' may be unachievable. Most religions have at their heart the idea that there might be ways to extend some form of participation, well beyond the allotted span. Part of what makes us human gains comfort in believing there is some continuity. And any religion which builds on that to create a better life this time around, must be considered of benefit.

Somewhere along the path to consciousness and intelligence, mankind developed a serious flaw of perception. There developed a tendency to take everything personally – the "Why me?" syndrome. When considering the human race as a whole, it must be remembered that we're playing a zero-sum game – there are an equal number of winners and losers. If you find yourself on the losing team during a particular game, you might try to identify some bias in the rules that worked to your disadvantage. The rules of nature have no bias whatsoever. Man, through various religious models, introduces a bias in the form of a God that supports one team in preference to another. The counter to this is quite simple – invent a new god that is on your side.

Many religious naysayers see the inability to shake hands directly with the 'Creator' as being a challenge to credibility. That is not necessarily a key requirement for benefits to flow to society or to an individual. We have very little understanding about the interaction of rules governing the physical and mental health of the human animal. Psychological well-being may play a much greater role in physical health than has been generally recognized until recently.

Some branches of belief subscribe to the idea that all life forms were created at virtually the same time, by divine intervention, and have followed a predetermined path to the situation that exists today – Creationism (ref: **Creationism**). This concept requires that every life, including our own, had a detailed blueprint from the start. There are some parallels between the rule-driven philosophy promoted in this book, and the concept of Creationist-blueprints, but a serious question arises about the level of detail that such blueprints might entail. Not only does Creationism pose a serious challenge to Darwin's theory-of-evolution, it is also at odds with the notion that mankind could be a chance result from the endless application of generic rules since time began.

Assuming that any particular individual deserves or receives special treatment ignores the premise that all laws were drafted long ago and apply equally to all parties. Taking the personal approach of a one-on-one session with the Creator may well have psychological benefits. Unfortunately, the guy next to you could be appealing for an outcome that is totally at odds with the one you have requested. If you both happen to be talking to the same Creator, then it is quite unlikely you both end up satisfied with the result. Perhaps a more realistic approach is to accept that the rules were originally drafted for a level playing field. Again, as a result of probabilistic outcomes along the way, the playing field could have become less level at some time. Inequality and inequity are inevitable because of the unstable equilibrium between the 'haves' and the 'have-nots'.

Conflicts often arise because of the different perspectives of Religion and Science. Much of this conflict might be explained by looking broadly at the models for each. Religious models tend to rely on a 'top-down' approach where each revelation is assessed to gain insight into its functionality and purpose. Results from this analysis are likely to raise many questions worthy of clarification, which then leads to further examination and discovery. Science employs a 'bottom-up' technique where observations and theories from many disciplines are compiled to create a knowledge base that forms the foundation from which to launch the next stage of discovery. Both of these processes have no clearly defined endpoint that might be interpreted as the ultimate answer.

In this book we have avoided addressing issues related to the origin of these magnificent rules of existence. Any discussion about creation belongs with a group of other philosophical questions that are doomed to remain unanswered. We have also tried to avoid cultural confrontation that inevitably arises when scientific understanding clashes with religious belief. However, there seems to be acceptance within all camps that something absolutely magical happened to deliver the experience in which we humans are deeply immersed.

Our existence is governed by an incredibly complex and beautiful set of rules. Does it really make a difference if they came from 'god', or were always in place? This argument will be philosopher-fodder until time ends.

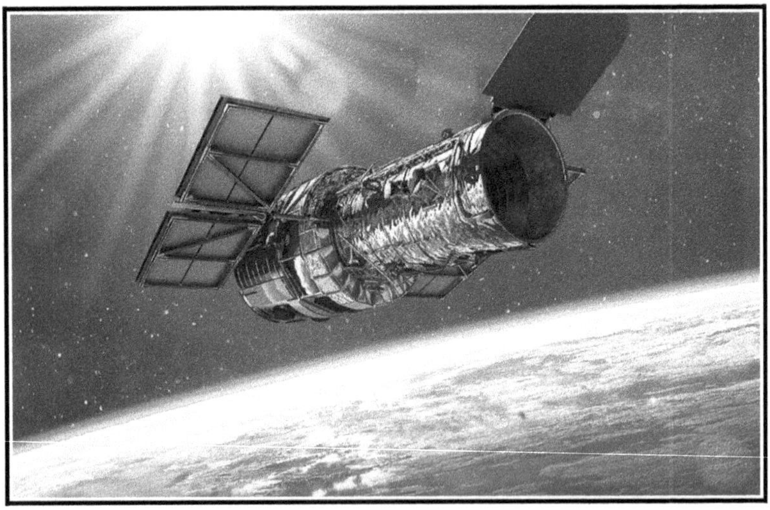

5. UNIVERSAL RULES AND PROPERTIES

"Our Universe must have been born into a framework of rules that may not be consistent with rules discernable from a human vantage point."

Perhaps the most basic way to understand what constitutes the Rules of Nature is to consider them to be descriptors for mechanisms in the various branches of science. It is then relatively easy to distinguish them from evolutionary characteristics that have possibly been optimized to suit a particular environment. For purposes of classification, inanimate objects are also assigned characteristics or properties. Material properties of such objects may vary as environmental conditions change, but evolutionary forces play no part in the derivation of those properties.

Rules of Nature are discovered by observing the various patterns of behaviour exhibited in response to various stimuli. Observable behaviour for all things in our Universe is controlled by three things – the fundamental rules of Nature; the environment; and any inherited characteristics or properties. At the top level, the rules specify the entire range of behaviour patterns available for each entity. The environment then sets bias towards behaviour in the rule range that is best suited for sustainable equilibrium within the environment.

In their most fundamental form, rules are prescriptions for behaviour of objects in the physical world. The Physical World extends far beyond the bounds of limited human perception. As such, it may not seem relevant to consider where these rules might be stored. However, exploration of the hypothetical concept of storage for the rules raises some intriguing questions. It could be logical to assume that the rules are an integral part of the space-time of our Universe, and contained within it. In relation to our Universe, the set of rules controlling our existence will be 'internal'. But there must also have been an external set of rules that directed the creation of the Universe and the prescribed behaviour of space-time within it.

Somewhere within the overarching master set of rules there must be a controlling mechanism to determine when a particular subset of rules should apply. For instance, when a bunch of molecules come together under specific conditions in a given environment, the rules kick in to dictate how they should chemically interact and combine. That will also establish a framework for the ultimate participation in a broader scenario. At whatever level we examine this, it does appear that the rules are external to the individual constituents. And that then prompts questions regarding the possible circumstances that might have led to the wondrous interconnected set of rules controlling all existence. In the bigger picture, if we accept the existence of external rules, our Universe must have been born into a framework of rules that may not necessarily be consistent with the natural rules observable from our limited vantage point.

At the highest level, the rules seem to be forever present, surrounding everything, just waiting for an opportunity to home in on a situation for which they have been designed. In the earliest stages of building our Universe, this might have involved elementary particles globbing together to create galaxies of stars, long before there was any opportunity for the evolution of life as we know it. As soon as conditions were right, life-forces came into effect. Taking our home planet as an example, around 4.5 billion years ago, objects orbiting around the Sun coalesced into a rocky planet within the Goldilocks zone of our Solar System. Fossil records have indicated that life started to appear on planet Earth within one billion years of its formation, and without any apparent trigger event to initiate the process (ref: **Abiogenesis**). There is no reason to suspect that this pattern for the emergence of life has not been replicated throughout the entire Universe.

Attached to many rules are sub-clauses, a sub-clause being an addition intended to increase comprehensiveness. It can often be quite difficult to identify these sub-clauses as separate from the underlying rule. In the same way that man-made rules have refinements that apply under certain circumstances, so too do the natural-rules. The big difference for man-made rules is that sub-clauses are usually added as a patch-up to cater for something unforeseen when the original rule was established. This is not the case for natural rules – they control the playing-field completely and there is no provision to alter them because of something unforeseen. Although we can influence parameters and environmental factors to produce various outcomes, we cannot directly control how any particular set of ingredients will combine and interact. How could we possibly imagine ever being able to fiddle the rules of Physics? In this regard, humans are purely observers. The stark reality is that we have no control in this area, and it is grossly egotistical to believe that we do.

All rules related to our existence belong to a group of intangibles, which also includes the fundamental elements Space, Time, and Gravity. These intangibles are invisible, but we know of their existence because of the influence they have on the behaviour of entities in the visible world. Before the General Theory of Relativity became accepted science, Space, Time, and Gravity were assumed to be independent, unrelated entities. Could it be possible that some future discovery reveals a mysterious relationship that binds the Rules more closely with other members of the intangible group?

Science is a discipline that primarily involves observation and modelling. 'Invention' often results when Applied Scientists and Engineers build creative solutions based on that modelling. Setting aside for the moment the more exotic models from some Theoretical Physicists, most scientific advances originate from careful analysis of empirical data. Ultimately, boundaries for human scientific discovery are set by the limits on what we are able to observe with the naked eye, or by using extended powers enabled by technology. Bounds are set at both the upper and lower levels of the spectrum. The structure of rules correlates broadly with our interpretation of existence, ranging from the cosmic heavens at the upper level, to the subatomic world at the base. At the lower extremity we might expect to find a base set of rules that are less likely to be separable then the more amalgamated rules at the higher extremity.

Cross-pollination is one of the benefits derived from attempts to clarify the domain of applicability of a particular rule of nature. The more we can identify and isolate sub-rules, the more opportunity there is to experiment and find where those rules may apply in different situations. We may discover something in a certain area that seems quite unrelated to another field of endeavour, and yet the benefit of applying something that is seemingly out of context could open new avenues of approach to many traditional problems. So the question has to be asked, "Is there any practical application for the categorization of rules?". Many of the high level natural-rules are quite obvious, as is the application of those rules. What is not quite as obvious is which sub-rules might be involved in the mix to contribute to a certain observable-outcome.

The rules we develop to describe the processes of nature are totally dependent upon the framework of observation. Several centuries ago, Sir Isaac Newton was observing gravitational effects on an apple. This led him to formulate a series of mathematical models that accurately quantify the laws of motion and gravity. Newtonian Mechanics quickly became adopted as accepted science, and remained unchallenged for about two hundred years. Unconstrained by conventional methodology that tied scientific modelling closely to empirical evidence, the legendary theoretical physicist Professor Albert Einstein introduced his Theory of Relativity. There was reluctance within the scientific community to immediately embrace the theory's radical concepts because they did not align with intuitive thought or experiences in the physical world. It has taken nearly one hundred years to empirically validate some aspects of Einstein's theories (ref: **Testing Relativity Theory**). What Einstein has demonstrated is that viewing the natural world from an entirely different, somewhat abstract perspective, can lead to revisions in previously formulated rules that may have earlier seemed complete.

Science is devoted to developing models that can explain physical observations and enable predictions that can be empirically tested. These models are closely aligned with the various rules of nature. When predictions from these models deviate from the empirical evidence, it becomes necessary to either discard the model, or refine it in ways that accommodate any newly discovered phenomena. The same is true of the rules assumed to be associated with these models, except that the rules are not modified – only our understanding and interpretation of them changes. A classic example of this is illustrated by the historical attempts to model the inner workings of atoms. From primitive beginnings that likened atoms to plum puddings composed of a positively-charged 'pudding' with embedded electron 'plums', current models of the atom incorporate various subatomic components, all controlled by their own set of rules. The implication here is that all rules are hierarchical, and we can never be sure of reaching the point where they can not be further refined.

Physical objects and materials are often categorized in terms of their quantifiable physical properties. These properties include Density, Strength, Coefficient of Expansion. Also included are less-quantifiable properties such as State (Solid, Liquid, Gaseous, etc.). The specific properties for a given object can vary quite markedly in response to changing environmental conditions such as Temperature and Pressure. The same physical entities can also have distinct Chemical Properties. In some cases, these chemical properties can be predicted from an intimate understanding of the internal atomic structures. Putting it another way – the Rules that appear to govern chemical reactions may not actually be independent, but instead derived from more fundamental Rules that apply at the atomic level.

Although chemical elements and compounds are formed strictly in accordance with the laws of atomic physics, they can exhibit unique behaviour well beyond anything that the rules associated with subatomic particles might suggest. For simple chemical interactions, predictability in outcomes from combining various elements comes largely from an understanding of the geometric configuration of the various atomic or molecular components.

There is a blurred distinction between 'rules' and 'properties', particularly so when considering various forms of life. What separates animate and inanimate entities is the ability of the animate to evolve and inherit characteristics and traits from their ancestors. A chunk of rock has a historical record that we can usually decode to discover where it's been and what it's been subjected to. The rock is composed of various atoms and molecules that have been assembled according to some high-level set of rules and circumstances, although that pattern of assembly will not have an influence on later rock formations. In contrast, all lifeforms have the ability to store historical records internally in DNA chains. This mechanism not only provides historical data, but also influences development in the generations to follow.

As various forms of life become identifiable as species, they develop characteristics or patterns of behaviour that are passed to future generations of the species. These characteristics have been selected from the set of possible outcomes permitted by the rules, to best suit prevailing environmental conditions. A particular characteristic might have originated as the probabilistic outcome from the application of some high-level rule, but once assigned to a species, it is no longer under the influence of the rule that created it.

Instinctual behaviour has frequently been identified in various animal species that does not appear to result from observation, training, or any environmental directives. This implies that instructions must already be in place prior to the birth of any animal exhibiting instinctual behaviour. Does it make sense to map this behaviour to the atomic level, and imagine possibly large assemblages of atoms carry with them instructions in preparation for the infinite variety of situations that might arise? Probably not. Animals are composed of billions of atoms, and the arrangement of those atoms in the form of DNA chains can contain behavioural directives derived from past experiences, that manifest as instinct. The important distinction here is that atoms and all subatomic particles behave strictly in accordance with the rules of Nature, or as we might refer to them in this context, the laws-of-physics. For complex multi-atom entities, there exists a possibility of an additional layer of instructions on top of the fundamental rules of nature.

For all biological species there is a vast amount of ancestral data that can be analysed to trace the evolutionary path that has led to present-day embodiments. Much of the historical record is stored within DNA strands of the species themselves. For non-biological entities, chemical composition provides limited information about the processes they might have undergone. Historic records, both internal and external, store the results of applying various rules, but not the rules themselves. The distinction between information, and the rules governing the collection and storage of that information, is a very difficult demarcation.

Dissecting processes of Nature and understanding the rules driving them, invariably leads to human innovation that results in quality-of-life improvements for many or even the majority of Earth's inhabitants. There is a wondrous beauty associated with these processes that is not always fully appreciated – they demonstrate an optimal solution for any given situation when all factors are taken into account. It is only logical that mankind should attempt to mimic these examples when innovating. Most inventions have roots that can be traced back to some natural processes.

A classic case of human invention mimicking Nature's designs is the way in which discovery of the aerofoil led to powered flight, and had a profound influence on many aspects of modern-day life. The Physics and Mathematics behind the design of the aerofoil are quite complex (ref: **Aerofoil-Lift**). And yet, long before subtleties of the mechanics were understood, careful study of the design of birds' wings produced practical prototypes enabling man to take flight.

The human brain is probably the most complex organ that has ever existed in a living species during the entire history of planet Earth. It is logical that mechanisms in the brain should be comprehensively probed to discover techniques that would benefit man-made computational systems. Huge advances in computer technology have occurred in the last few decades, but there is a long way to go before they can match the sheer efficiency of an organic brain to process, store, and retrieve information. The brain still holds many secrets that we are as yet not able to unlock. One of the most intriguing of these secrets is how vast amounts of image data can be so densely compacted and yet so easily retrieved. Computational power within an organic brain is truly amazing, but computer scientists are rapidly closing the gap between Nature's solution and what can be achieved artificially. Recent advances in Quantum computing indicate that dramatic increases in computer power may not be far off. This may turn out to be a rare instance where human ingenuity comes close to achieving the solution delivered by Nature.

Organic structures have a natural advantage with their ability to construct in three dimensions. Manmade computer fabrication currently relies on stacked, single plane components. Perhaps future computer technologies can align more closely with the example from nature and incorporate full three-dimensional interconnections. But 'structure' is only half of the problem. Computer systems are built from basically two components – 'hardware' and 'software'. Once we figure out how to construct three-dimensional core components, the next challenge will be to design algorithms and write programs that can capitalize on the new framework. Not a trivial exercise. It might appear at first glance that the brain could be seriously disadvantaged because of a lack of regular software upgrades and patches. However, this is yet another reminder that the algorithmic rules of Nature do not change, and do not need to change if they have already been optimized for the most efficient outcomes.

However, Nature may not always have been able to anticipate human ingenuity. One advantage humans have in the inventive process is a unique ability to assemble things. Most modern inventions today rely heavily on a complicated assembly of components drawn from many diverse sources. Take as a starting point what might be arguably the most uniquely human invention – the wheel. The humble wheel set the foundation for the huge variety of mechanical devices that permeate every corner of civilization. So how could it be that the wonderful wheel is missing from the natural world? Non-biological naturally occurring processes are randomly chaotic and rarely produce structures with the high degree of symmetry required for a functioning wheel. Biological specimens, particularly animals with legs, have a degree of symmetry, but there is a reason why limitless rotation is not possible in one section of a living organism while another section is stationary. A requirement for all sections of the single living organism is that they maintain continuous linkage.

Extending this idea to an interesting, and perhaps controversial conclusion, it may be possible to assemble a bunch of components that will attract a set of rules to become what we understand as life. As long as we follow the rules in providing the right components in the right environment, then the rules should apply equally well whether the situation was contrived by man, or whether it happened from natural causes. Scientists and biologists are moving ever closer to proving that life could be created using relatively simple molecular components and structures that lack life-force. (ref: **Synthetic Cell Growth**).

There are two ingredients essential for the creation of recognizable life that are near impossible to emulate in a laboratory environment – Time and Scale. Evolution by natural selection is a slow probabilistic process that is heavily dependent on both these factors. Although researchers might be able to commence an experiment to create new forms of life, the results from such an experiment may not be available until long after all traces of humankind are consigned to the fossil record. However, creation of viable life is not the purpose of this exercise. If it can be demonstrated that biological functionality can result purely from interactions between chemical elements which commonly occur throughout the Universe, then it follows that life itself must also exist in abundance beyond our planet.

Unlike the Rules that exist forever without any beginning or end, all forms of life have a limited and clearly defined lifespan. This lifespan is different for every form of life, and might be considered a characteristic belonging specifically to each plant or animal species. In addition to the limited lifespan of individual members of a species, the species itself is also destined to become extinct, although the timing of that extinction is less well-defined. One consequence is that we need to face the uncomfortable reality regarding the mortality of human civilization.

There is no shortage of doomsday scenarios which predict, with varying degrees of likelihood, the extinction of the human race. Superbugs; Nuclear holocaust; Meteorite strike; Overpopulation; Scarcity of resources, and of late, Global Warming and Artificial Intelligence. Whatever the cause, at some point our species will hand domination to another, perhaps with a brief pause while the Earth reconfigures to make the environment amenable to the next iteration of life forms.

One way of looking at this is that everything in nature has a specific timeline. On a cosmic scale, stars take billions of years to form, and maybe another few billion years to burn out. Closer to home, us humans can optimistically expect our personal timeline to extend out to approximately 100 years. The humble butterfly can count a few days during its lifetime. For a pi-meson in the Earth's upper atmosphere, the half-life is measured in nanoseconds. All these timelines are intricately interwoven to play a part in the beautiful tapestry of existence. Contemplating the end of human civilization could be nothing more than an acceptance that our time is up; an admission that we're just another animal species governed by the same set of rules for sustainability as all others.

As humankind has developed intelligence, knowledge, and the ability to upset sustainable equilibrium on a planet-wide scale, it has tended to all too often move down the path to damage and destruction. If this is an inevitable trajectory for a species as it gains intelligence, it would be logical to expect the rules would restrict the amount of damage that could be done on a cosmic scale. Perhaps the limited duration for an intelligent civilization, taken together with the astronomical distances involved, constitutes a natural safety factor built into the rules to guarantee that no intelligent species could ever attain enough power to upset anything outside their immediate sphere of influence. In fact, this isolation has sand-boxed the human experiment. If humans had any chance of influencing things on a cosmic scale, they would have had to act very soon after the Big Bang. There again it appears a safety-valve prevents this from happening. By the time any intelligent species has evolved, it would be too far distant from ground-zero to have universal consequences.

There is another safety-factor subtly built into the biological rules for intelligent life – a slow evolutionary process. The long delay involved in the development of higher-intelligence creatures discourages humans from conducting experiments aimed at creating lifeforms from primitive organic ingredients. Moral and ethical considerations aside, it is more than likely that somewhere in the world organic manipulation will be pushed to alarming levels. We can only speculate about how the rules formulated by nature might apply in such circumstances.

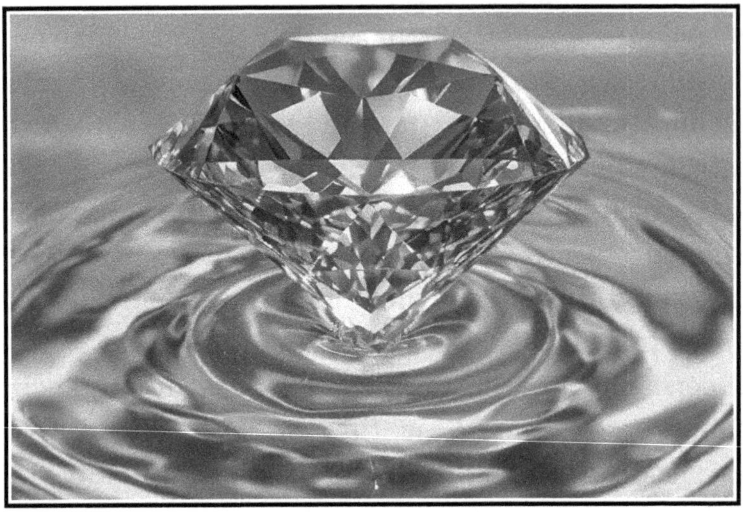

6. IMMUTABILITY AND INSTABILITY

The quest to identify and decipher the Rules of Nature is endless, with the joy of the journey far outweighing any frustration at never being able to attain a comprehensive understanding of existence.

One of the most significant and far-reaching scientific observations has been that, within the limits of current engineering prowess, no variation in the Rules of Nature has ever been detected. This immutability has profound implications for the way we should think about the rules. For any Rule to apply during the entire existence of our Universe, it must have been designed with intimate advanced knowledge of all future scenarios. Not only that – it must also take into account all possible unintended consequences of applying that rule in combination with an endless selection of other rules.

Suppose for a moment that the rules are derived from an iterative trial-and-error process. Obviously this evolutionary experiment must have been completed long before the birth of our Universe. Not only does that clearly imply that there were processes at play long before the Big Bang, but also begs the question as to how long these experimental rules evolved before they were ready for prime time in our Universe. The notion of experimenting with the rules becomes quite absurd, and it is not a giant leap of faith to accept that the rules that apply to our Universe have remained unchanged for eternity.

In order to appreciate the mind-boggling complexities involved in establishing rules that must stay relevant for billions of years, it might be helpful to assess the design process as it applies in a human context. Human designs and inventions quite often have a very short lifespan. The reason for this is twofold. Technological advancement can render devices obsolete in a very short time. In addition, a changing environment may mean that the functionality provided by a device is no longer relevant. To create devices with the best chance for a reasonably long life requires a full understanding of the technology of the day, plus careful assessment of any influencing technological and environmental changes that are likely to occur in the near future.

One of the most incredibly successful designs in modern times has been the Voyager spacecraft. After nearly 50 years travelling to the outer reaches of our Solar System and beyond, these machines are still capable of delivering the functionality for which they were originally designed (ref: **Voyager Spacecraft**). Because of the ever-increasing pace of technological advancement, future designs and inventions are likely to have even tighter constraints on functional lifespan.

In a bygone era, human innovation was derived from direct everyday experience. The whole concept of 'Time' was thrown into question when Einstein revealed that we could no longer depend on a steady rate of the passing of time. Around 130 years ago when many scientists believed they had nearly all the answers, it was a comfort to believe that humans were part of a process that was predictably sliding along the timescale. Although Einstein's contributions were of monumental scientific significance, they did not emerge in isolation. They were fostered by a maturing technical landscape, and a scientific community more willing to accept well-reasoned radical postulations. Thanks to Einstein's efforts, scientific thinkers were emboldened as never before.

Mathematicians are far more comfortable when dealing with Infinity than are their colleagues in other branches of science. Nothing in our observable world gives any clues to understanding infinite space and eternal time. Into this mix of the unfathomable, we must now include an endless set of interconnected rules. The possibility that any of Nature's rules might be experimentally tweaked without negative consequences, becomes even more remote.

It would be a mistake to believe that "immutable" translates to "inflexible". For the application of any specific rule, there is a set of possible outcomes with associated differing likelihoods of occurrence dependent on the set of parameters. No matter how weird or unlikely the result may seem, the original rule was defined long ago to accommodate the possibility that such a result might occur. Diverse outcomes are not due to any variations in the rules themselves, but are produced by the application of rules that permit non-definitive results.

Do not necessarily expect the application of a given rule to always produce precisely the same result under the same conditions. Most rules in nature have a probability of delivering an outcome within some predefined range, but any outcome will also depend upon influences from environmental parameters as well as a multitude of overlapping rules. Some rules result in an outcome that is inherently unstable. This instability can also contribute to what we ultimately measure as probability of occurrence.

Until an event or circumstance is observed in nature, the possibility that such an event could occur is pure conjecture. When an event occurs multiple times under the same environmental conditions, it is possible to estimate the probability of future repeat occurrences. For events resulting from completely random behaviour, the average probability of occurrence can be calculated for the events overall, but the value will not necessarily align with the probability measured for a large-scale collection of events in a different sample. If the average probability of a particular outcome can be accurately reproduced across multiple independent situations, it would be fair to assume that the probabilistic outcomes are tightly controlled by one of the many rules of nature.

It might seem logical that evolutionary processes dedicated to delivering optimal solutions should produce identical results in any situation where the environmental conditions are identical. However, the natural selection on which evolution depends is not a choice made from a boundless set of options. Candidates for selection all result from random mutational events, but those mutations must fall within the range of possibilities allowed by the rules of biology.

As humans, we owe our very existence to the diversity of outcomes linked with application of the rules. Biological evolution and diversification are totally dependent on trial-and-error processes to determine the best fit within a given set of environmental parameters. Every rule has within it defined ranges of permitted outcomes, but we cannot guarantee that application of a particular rule will deliver a specific outcome.

The evolutionary process of Natural Selection breaks down if there are no distinct choices available to make a selection from. Those choices are provided from randomly generated mutations. There are many sources capable of causing mutation in living cells. To allow time to evaluate whether a particular mutant variation will deliver an advantage to future generations, the total rate of mutation due to all sources must be suitably limited.

The detrimental effects on life from exposure to both particle and electromagnetic radiation are well established. Mother Earth has two mechanisms to limit exposure to both types of radiation. Charged particles from the sun are deflected by Earth's magnetic field, and a relatively thick atmospheric layer absorbs much of the harmful electromagnetic radiation. There are some places on Earth where radiation does not penetrate – the dark world deep below the ocean's surface. And yet, even in the most inhospitable environment imaginable, evolutionary forces are at work as evidenced by the existence of primitive lifeforms surviving around deep sea vents.

These deep sea creatures are not just mere remnants of life that have drifted down to the depths. They have demonstrated a remarkable ability to adapt for survival in conditions quite toxic to all other forms of life on Earth. The existence of these strange creatures with the ability to evolve and survive without assistance from externally induced mutation, suggests inbuilt rules might allow a natural degree of mutation or error rate in the generational replication processes of cellular life.

Once evolutionary forces have run their course and a particular species is perfectly tuned to its surroundings, does the evolution of that species come to an end? This would be akin to asking what more is there to learn when we eventually know everything. Evolution may not necessarily end once a species is in perfect balance with its surroundings. Even in the most harmonious situations, there may still be opportunities for slight improvements in efficiency – for example, reduced energy requirements to produce the same result.

Intelligence is a consequence of increased brain capacity, and it is the main factor driving future refinements in efficiency.

As animal lifeforms move higher up the intellectual spectrum, a truly spectacular transformation takes place. For simple lifeforms, the developmental path follows from random mutations and natural selection. Once species reach a certain intellectual threshold, they are enabled and emboldened to conduct their own experiments, with serious implications for future evolution. Higher intelligence creatures possess an incredibly powerful tool – "imagination". They no longer need to wait for beneficial outcomes to be delivered by chance. The inherently slow natural process of trial and error is greatly speeded up. The irony here is that this burst of intelligence may actually be signalling the impending extinction of a species which has gained too much control over its own destiny.

There might be a downside to being a highly intelligent species. Although individual members of the species may have relatively long life spans, the timeline to extinction of the species as a whole, could be relatively short. Microbial life has existed on earth for many millions of years. It was able to do so by slotting seamlessly into the ecosystem without demanding additional accommodation, and often contributing an overall benefit. Humans have existed in their current form for less than a hundred thousand years. The stresses currently placed on the planet's environment could indicate that the harmonious balance essential for sustainability no longer exists between the human species and their home planet.

Planet Earth has produced an endless diversity of plant and animal species. And that has been possible because of the vast variety of environments the planet supports. The rules that govern all forms of life were in place when the Earth was a seething mass of volcanic activity and also during various Ice Ages. A set of rules with the ability to combine and coordinate to produce a blueprint for dinosaurs was always present. The blueprints never actually existed, although, from the way dinosaurs were so perfectly suited to the environment of the time, it might easily be construed that they were designed well in advance of their appearance on planet Earth. There is an infinite variety of possible life-forms that we will never discover because the necessary incubating parameters are inconsistent with those of our own environment.

Dinosaurs, and their failure to return again after becoming extinct, offer an excellent example of how rules can produce vastly different outcomes, even though the environmental conditions might remain relatively unchanged. There seems to be universal agreement that the dinosaurs became extinct because of the dramatic environmental change that occurred as a result of Earth's collision with a massive meteorite. At the height of their reign, various dinosaur species dominated life on Earth. However, when the planet returned to a stable condition potentially capable of supporting dinosaur life, they did not evolve as they had in the past. One plausible explanation is that the meteorite strike and associated volcanic activity caused dramatic changes to the atmosphere. This in turn produced subtle, long-lasting changes in the plant food-chain that was a critical factor in determining the path of dinosaur evolution.

Given the probabilistic nature of destiny, what can we infer about life throughout the Universe based purely on data sampled from a single life-supporting planet? The short answer is, actually quite a lot. Life on Earth underwent a major reboot at the end of the dinosaur era. This provided an excellent opportunity to study two separate streams of evolution in a similar environment. Flightless, land-based animals in both situations evolved to have many features in common.

- Dual eyes and ears
- Four appendages
- Robust circulatory systems
- Large brains
- A high degree of bilateral symmetry

The chance of finding planets elsewhere in the cosmos, with exactly the same conditions conducive to life as on Earth during its habitable periods, is extremely remote. Planets will most likely be larger or smaller, hotter or colder, wetter or dryer, have different atmospheres, and have a vastly different mix of mineral and organic resources. However, since the rules governing the formation of life are universal, it would not be unreasonable to expect alien life to develop many of the features that characterize life here on Earth. Although there are no guarantees of finding another Beethoven, Einstein, or Shakespeare, there are bound to be intellectual equivalents in any highly developed alien civilization.

There is one observation about life on Earth that may well apply throughout the entire Universe – Life is pervasive in the extreme, and as long as a few fundamental elements are available, it would seem reasonable to expect that life in some form might evolve. There have been endless forks along the path of human evolution. If the option at any one of these forks was selected differently, the outcome for human civilization could be radically different from what exists today. One interesting alternative fork to consider is what might have happened if dinosaurs continued their reign as the dominant species. Would they have continued evolving and developed higher intelligence, or might they have surrendered their dominant-species crown to some smarter, smaller, more agile creatures? The possibility exists that humans might still have become the dominant species, even without the untimely extinction of dinosaurs. There is a danger that we could arrogantly assume that the human animal is the ultimate manifestation of intelligent life.

While we can readily research the logic and paths of evolutionary-biology, a similar degree of examination of physics at the subatomic level provides a daunting challenge to both Engineers and Theoretical Physicists.

The broad category of rules known as the laws-of-physics have historically relied on observational science and discovery. These may have seemed to be 'absolute' rules that could not be distorted by some undiscovered sub-rule. Increasingly we are discovering that even the most fundamental rules governing Space, Time, Gravity, etc., are far more interconnected than once assumed.

Probability and Uncertainty are fundamental elements of Quantum Mechanics. This is a relatively new scientific discipline that introduces concepts more foreign to our everyday experiences than we could ever have imagined. Quantum Entanglement is an excellent example of unexplained phenomena that has challenged the world's greatest minds, and yet we still have no idea how it is even possible. Incredible advances in engineering allow us to see what we cannot intuitively believe or observe with naked eyes. The age has passed where we could accept as valid only that which might be revealed by our unaided human senses. That said, it should not be taken as free licence to totally divorce predictions of future science from that which has been established today.

Natural mutation of cellular life has an equivalent in the atomic world. Many heavy atoms suffer an inherent instability that can be detected as radioactive decay. For radioactive atoms, half-life is a measure of probability that an atom will split into component elements within a certain period. Even though the timing of decay for a particular atom is quite unpredictable, the half-life for a large sample of atoms taken as a whole is precise and reproducibly constant. So it would seem there is an internal rule associated with each radioactive element that specifies the exact rate of decay.

The mix and distribution of elements throughout the Universe is well explained by the model from current science where hydrogen atoms are fused together in star-based thermonuclear reactions to form heavier and rarer elements. The process is completely random and utterly chaotic. What is not random is the proportion of isotopes included in the mix. Many elements have isotopic variations, but when present, isotopes constitute an exact proportion of any large-scale sample of a specific element. For example, hydrogen isotopes occur naturally in the form of deuterium and tritium; uranium has several isotopes that are lighter or lighter than the more common U-238. Any sample of naturally occurring hydrogen will contain approximately 0.02 percent deuterium and tritium isotopes. Once again, this illustrates the distinction between chaos and probabilistically reproducible behaviour.

On a cosmic scale, there are many tantalizing discrepancies between current scientific model predictions and increasingly accurate measurements. There is fertile ground at both the cosmic and subatomic level to discover complex sub-rule extensions to some of the more basic rules gleaned from direct observation. Dark Matter and Dark Energy are currently nebulous entities that bundle a vast trove of sub-rules which will doubtless provoke a rethink for many of the rules already identified by science. When we better understand the contribution these make to the stability of our existence, it might be possible to make a more realistic prediction about the ultimate end of the Universe.

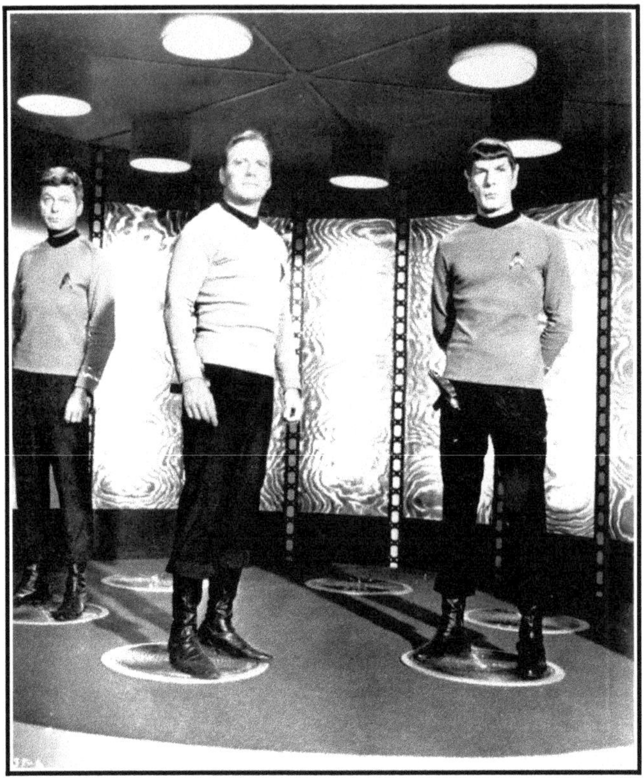

7. SCIENCE FICTION – A REALITY CHECK

The Star Trek / Doctor Who influence –
"Beam me up Scottie" – I don't think so!!

Science Fiction covers the entire credibility-spectrum, from almost plausible through to virtually impossible. What makes it so tantalizing is that we're never quite sure where on this spectrum a particular variant might belong. To achieve some credibility, the storyline must at least have tenuous links with the accepted science of the day. But credibility is not necessarily a requirement for good Science Fiction. With these thoughts in mind, let's explore some of the more improbable concepts.

The fact that there is no definitive outcome from many of the natural rules creates a real problem when we think of some of the more popular ideas from science-fiction. One of the greatest challenges is faced by the Star Trek invention of 'Beaming Up' – a wonderfully appealing concept of being able to ignore all Physics related to space-time management, and permitting an instant transfer of physical material from one place to another. There is currently some extremely fascinating investigation being done to explore the mysterious mechanism known as Quantum Entanglement (ref:

Quantum Entanglement). It is becoming increasingly likely that we may be able to replicate the properties of a single particle at a distance. However, there is a huge leap-of-faith in assuming that we might ever have technology capable of transmitting roughly 10^{28} atoms in the average human to another place. Once again, the characteristic imprecision of many of Nature's rules is likely to come into play. Only a few feral atoms in the transfer could lead to a catastrophic result. The chance of surviving such a transfer in a recognizable state is unlikely in the extreme. It is interesting to note that, in a nod to the realities of physics, seat-belts come as standard equipment on the latest Starships.

Another intriguing idea from science-fiction is to imagine that each grain of sand in our Universe may be another self-contained universe in its own right. We will never have the ability to burrow down deeply enough to validate this theory. But we certainly can investigate what regulation and set-of-rules might be required to allow such a situation to be possible. Imagine the trauma in one of those minute universes if it was unfortunate enough to become attached to the ball in a reasonably violent game of beach-volleyball. The inertial-forces alone would wipe out any possibility of life as we might recognize it. However, we have been privileged only to a very myopic view of all the wonders that Nature can offer, and we cannot with confidence exclude things that may seem outlandish. It might be conceivable that rules which apply on a very minute scale include a mechanism to protect sub-universes from the devastating forces of Inertia. Pushing the bounds of speculation, perhaps there exists some kind of inertial shield that has not yet been discovered. It may even be protecting our current Universe from destructive outer-forces. However, the real challenge to the credibility of this concept is that it defies all the prevailing laws-of-physics. It implies that the laws are relatively independent of scale, whereas there is a stark difference in what we observe as quantum physics at small scale and astrophysics at the high end.

We can never be certain about what happens beyond our event horizon. It could be that our Universe is in the midst of a volleyball game and, from the 'inside', we have no way of detecting it. Investigating subatomic entities at the lowest level on the scale spectrum reveals there is far more waiting to be discovered than we could possibly imagine. Similarly, looking out to the heavens presents an infinite depth to be explored. Our relative position on the scale spectrum lies somewhere between the extremities, but the exact position cannot be determined since the end-points will forever remain unknown. It is possible that scientists and engineers are very close to finding the smallest subatomic particle that forms the basis of all other particles and cannot be subdivided further. There is no way we can ever be confident that point has been reached. Brilliant engineers will forever continue their quest to dig deeper into the mysterious world of the minute, but there is no flag to indicate that the quest is over.

The boundary between science and science-fiction is determined by two factors – an incomplete knowledge regarding the laws-of-physics, and the practical limitations imposed on human-engineered solutions. Flaws inherent in all human constructs will ultimately dash all hopes of colonizing another planet, even a close one. And that is quite apart from the complete inability to justify the truly vast financial expenditure involved. The Space Shuttle disaster brought home the realities of engineering fragility.

But engineering issues are far from the most significant challenges faced when considering the establishment of a base for humans on another planet. Biology is the show-stopper. What is often overlooked is the complex evolutionary path humans took to become the dominant species on Earth. There was no predefined blueprint for the design of humans that was scheduled for implementation once the conditions were favourable. Over millions of years, the human species has been fine-tuned to become a functional fit with conditions on Earth. One reason for that very slow development is the relatively long period from one generation to the next – around 30 years for humans. Many tiny creatures have life-spans no greater than a few days. The resulting short generational cycle increases the opportunity for mutational variations that drive evolutionary development. To transpose humans to an environment for which they were not designed is no mean feat, and any chance of success is extremely low.

Colonization of another planet would require maintaining, for many thousands of years, an artificial environment mimicking Earth's. Evolution would simply be too slow to allow humans to adapt to any unaccommodating conditions in an alien world. If expense and effort are no object, many of the elements required to maintain an artificial environment on another planet are possible using today's technology. In that sense, science-fiction tales related to human colonization of other planets, do not seem outside the realms of possibility.

Because humans have evolved slowly to function effectively on planet Earth, it is easy to overlook the environmental subtleties that we take for granted. Any attempt to colonize another planet would require a huge effort to continuously simulate conditions here on Earth. Alien worlds may not be surrounded by a magnetic shield that protects us Earthlings from harmful radiation. The air we breathe must be a precise mix of various elements for humans to function normally and healthily. Humans have evolved in response to Earth's available nutritional elements. This inevitably led to a dependence on those elements for survival. A planetary outpost hoping to sustain a permanent human presence would also need to provide an uninterrupted supply of the same mix of nutrients. The list goes on, and the challenges for creating a sustainable artificial environment on a foreign world become ever more daunting.

With the best will in the world, there are some aspects of an alien environment that may be difficult, if not impossible to control. Gravity has a huge influence on the form-factor of all living species, both plant and animal. Studies of astronauts who have returned from lengthy stays in the weightlessness of Space have revealed many subtle effects that gravity has on the human metabolism (ref: **Effects of Weightlessness**). For long-haul journeys in Space, gravity can be simulated in spacecraft by maintaining an appropriate degree of rotation. Centrifugal forces would substitute for the natural gravitational forces that apply on the surface of the Earth. Other techniques for producing artificial gravity belong strictly in the science fiction category until given credibility by some new discovery in physics.

Even if the human animal was able to suitably adjust to conditions on a different planet, we would not be evolving in isolation. Every human being is sustained by a microcosm of microbial life forming a complex symbiotic relationship. Around half the cells in the average human are not actually human at all. And the non-human cells might have different ideas about how they might react to a new set of underlying environmental parameters. Because they have a much shorter evolutionary cycle, microorganisms can react quickly to changes in environment. There are no guarantees that their future development path will require them to remain in harmonious synchronization with the human species.

Some of the most far-reaching decisions regarding the establishment of an off-Earth colony, must be made at the outset. Obviously, the first few 'boatloads' of colonists will contain a diverse genetic mix selected from various human tribes. But what other forms of life should be included? This is an exceedingly difficult question to answer. Every form of life we know about is part of an integrated ecosystem that involves other lifeforms, both higher and lower on the food-chain. How would it be possible to take a selection of isolated lifeforms and expect them to survive and thrive in a hostile environment? If we decide to involve only humans in colonization, then every individual headed to the new world must be sterilized to remove all forms of life that might be considered non-human. And this is where the dream comes unstuck – we cannot separate and isolate humans from all the intricate biological mechanisms essential for survival.

Contrary to all common-sense, let's suppose a decision was made to include a few pet species in the initial mix with humans on a colonization mission. Questions then arise as to the minimum number required for each species. We could follow the strategy adopted by Noah for his Ark, but that introduces the undesirable effects of inbreeding. In addition, careful studies would be needed to determine the feasibility of recreating the food-chain environment surrounding every introduced species. The more we extend this concept, the more preposterous it becomes. No animal or plant species on Earth evolved in isolation – their development has been totally dependent on external forces surrounding them as other species evolved and matured. To pluck a species from the harmonious environment it has adapted to will doubtless result in adverse or even fatal consequences. All these factors serve as a reminder of just how neatly we humans sit within the cosy confines provided by Mother Earth.

The notion of intelligent lifeforms visiting Earth from the far reaches of outer space has long been a theme for Science Fiction writers. However remote the place those visitors call home, their incubation and evolution was controlled by the same identical set of rules that led to the existence of humans. It follows that they might have many of the same features and frailties that human beings do. Conversely, we might develop characteristics similar to theirs if our civilization survives long enough. This train of logic probably explains why the majority of science-fiction aliens are humanoid in form.

An intriguing mystery to investigate, even hypothetically, is how alien life might have acquired intellectual superiority. Were they originally like us and evolved a lot longer? Or was it some magic chemical compound they discovered to enhance intelligence? Or maybe it was some key aspect of their environment that was lacking on Earth? Another alarming possibility is that there could be self-inflicted factors here on Earth that might actually be restricting intellectual growth in humans. The Industrial Age has introduced pollutants to the atmosphere that are detrimental to ecological equilibrium.

Who among us would not be excited to encounter 'little green fellas' who have travelled for billions of years searching for signs of life alien to them. They might be exhausted from confinement during the long journey, even if they were hibernating in stasis. Having discovered us, the scouting party might be keen to send a message back to home base advising they had found a great place to explore. Problem is, home base would have long since disintegrated and quite possibly currently resides within a Black Hole. The astronomical distances and huge travel times suggest the near impossibility that two civilizations from different star systems might ever be able to interact.

Even if the absurdly improbable scenario occurs as described above, Earth may not be the ideal place for an alien race to contemplate colonization. The planet is already stressed by over-population, which places demands on natural resources beyond the limits of sustainability. Presumably this alien species would be technically superior and much smarter than ourselves? They would be acutely aware of the perils of attempting to integrate with a pervasive population that is already well established. Popular science fiction would have us believe that they might choose to live out their remaining years playing hide-and-seek with humans, only revealing themselves to a select few. This may be preferable to the alternative of embarking on another billion-year journey to find the next outpost in the galaxy that could be interesting.

Time travel to the future has always been an exciting prospect for science-fiction fans. However, the appeal of such a journey is dampened with the realization that it would be a one-way trip. Time-travel in a forward direction on a personal level is scientifically feasible and relatively easy to accomplish. The only requirement is for an individual to be put to sleep, with all metabolic processes slowed or on hold. Biology has not quite achieved this degree of control over metabolism, but there is no reason to believe it will not be possible in the not-too-distant future.

Travelling backwards in time is a far more challenging task. To fully understand why this is virtually impossible, it is necessary to delve into the definitions of time and appreciate that there are actually two distinct variations – Space-Time and Sequential Time. We have Einstein to thank for quantifying the curious relationship between space and time in our Universe. In the context of Space-Time, time is the measurement of the period between any two event points. It can range from zero to the largest positive number imaginable, but cannot be negative. Sequential Time is not concerned about the period between events, but rather the order in which events occur. When Sequential Time is zero, it means that events at the two points of reference occur simultaneously. Otherwise, the Sequential Time between two events is either 'before' or 'after'.

Commentators, even some within the scientific community, have claimed that nothing could have existed before the Big Bang because time itself did not exist.

Quite possibly the Space-Time of our Universe had no meaning prior to the Big Bang. That has absolutely no bearing on events that may have occurred in Sequential Time before the Big Bang, and certainly does not preclude the possibility of their existence. It is also possible that our Universe was born within a higher-level Space-Time framework.

The more that scientific discovery reveals, the better we can identify the division between the possible and the impossible. As long as our understanding of the laws-of-physics and the rules of nature is incomplete (which is likely to be forever), we cannot categorically state that an idea from science-fiction is impossible. The best we can do is judge how improbable it might be in terms of currently accepted science. There are many untouched areas where science has not yet ventured. One such area is telepathic transfer, and because of the lack of scientific evidence, either for or against, this topic is ripe for speculation among science-fictionalists.

There is considerable evidence suggesting some mysterious form of communication involving brains which does not utilize the known human senses. Unfortunately, the bulk of evidence for telepathic transfer is anecdotal, and can be only too readily dismissed as circumstantial. Any mention of the paranormal is guaranteed to produce audience polarization between believers and sceptics. What is often overlooked is that there could be a perfectly rational scientific explanation. To date, science has not yet identified a mechanism capable of directly exchanging information between human brains. And there is a simple reason why investigation in this area is so slow to produce useful and conclusive results. The process of turning observation into scientific hypothesis or ultimately laws, depends upon results that can be reproduced under controlled conditions. For paranormal or psychic events, it is very difficult to recreate the circumstances that might have been necessary to facilitate the observed phenomena. Without doubt, there will be rules related to these phenomena that still await discovery.

So what mechanisms are available from the established laws-of-physics that might explain mental telepathy? A prime candidate is some form of biological Quantum Entanglement. This little-understood "spooky action at a distance" could provide an explanation for the anecdotal evidence of information transfer between seemingly unconnected entities. Another possible explanation, and one that may be a lot easier to validate empirically, is that for mental exchanges between participants in close proximity, very weak magnetic fields produced by the brain may provide the mechanism for information transfer.

Much has been written of late about the threat to human civilization from Artificial Intelligence. The threat is misconceived, and the development of computational algorithms applied to massive databases was foreshadowed in the early days of computing. As a result of ever-increasing connectivity, there is a danger that malevolent actors could disrupt essential services, but that is not related to the notion that autonomous robots powered by artificial intelligence could take over the world. The real challenge required for robots to gain autonomy is some reliable self-repair mechanism. Self-repairing systems designed by humans have traditionally relied upon redundancy. In addition to built-in redundancy, regenerative processes are often used by biological entities to achieve an optimal lifespan for a given species. Note that biological self-repair could continue indefinitely, but that might be at the expense of normal evolutionary processes. Autonomous robotic machines require substantial, resource-hungry ecosystems devoted to production and maintenance, and it is difficult to imagine how such facilities could be developed without a huge commitment from complicit humans.

Sometimes science-fiction writers produce very accurate predictions of near-future technology. Around 80 years ago, the comic-book detective Dick Tracy introduced the two-way wrist radio. What makes this most remarkable is that it predated the invention of the transistor by several years, and was decades before integrated circuits enabled the level of electronic miniaturization that we see today. Electronics developed naturally along a pathway that led to smaller and smarter devices. However, even with these advances, there is one feature of the wrist radio that may remain in the realms of science fiction for a long time to come. Physics places a restriction on the volume of low-frequency sound waves that can be generated by small devices. As is often the case, we may find clues in the natural world that provide an efficient solution.

In this chapter we examined some of the concepts proposed by science fiction writers. The distinction between good science fiction and postulation from eminent theoretical physicists is quite blurred. The next chapter investigates some aspects of theoretical physics where competing propositions are queued awaiting empirical evidence to allow them to move into the realms of accepted science.

RULES FOR ETERNITY

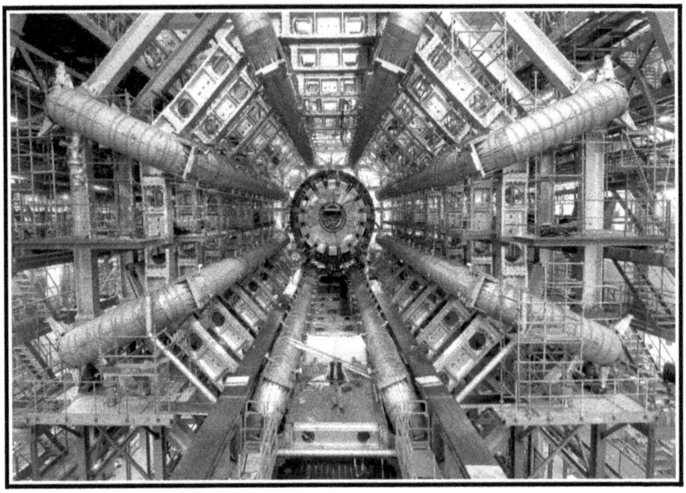

8. A BIG ASK FOR FUTURE SCIENCE

"One of the biggest problems in predicting the future is that you might be in it before you realize."

Broadly speaking, scientific endeavour can be divided into two categories – the discovery stage and the application phase. Exploration to discover and validate new laws-of-physics is becoming prohibitively expensive, and requires planning and commitment for a decade or more. In a sense, the low fruit have already been picked. Applied Science is a field of human endeavour much closer to home that is capable of delivering almost immediate results. Significant knowledge of the laws-of-physics has been accumulated, and much of this knowledge has yet to be deployed in practical applications. An intimate understanding of these rules leads to innovative solutions that can produce practical and near-term benefits for humankind.

There was a watershed moment in the history of science where postulation went well outside the frame of the observable. Traditionally, models were proposed to explain tangible aspects of our experiences and interactions with the natural world. Albert Einstein broke with that tradition, and was not shy of presenting scientific concepts and models that ran counter to everyday experience. More than a hundred years after the publication of his theories, some aspects of his work are still being empirically confirmed.

Early on, scientists hoped to find a single all-embracing model that would explain everything in existence, from the smallest components of matter to the outerreaches of the Cosmos. Over time, it became obvious that the rules developed from everyday experience were insufficient to explain the inner workings of atoms, or the big-picture phenomena driving cosmic processes. New areas of physics were developed to deal with the anomalies. Quantum Mechanics is a good example of 'new' physics that was created to explain behaviour in the mysterious subatomic world. The idea of a unified theory of everything is back in favour, but this time with a different slant. No longer is human experience taken as the baseline. Instead, the focus is on investigating ways in which our experience is influenced by fundamental phenomena. Constraints on human powers of perception and investigation have prevented some rules from being fully elucidated, and their range of application misinterpreted. The Holy Grail for scientists is a unified theory that marries quantum mechanics with traditional physics models.

The best indicator to highlight gaps in scientific knowledge is when currently accepted models fail to match the empirical evidence. If we already had models to fully explain everything, there would be little incentive to search beyond the horizon for possible new frontiers. There are many instances where current scientific models fall short. In the early part of this chapter we examine some of them, and speculate about whether or not those knowledge gaps are likely to be filled in the foreseeable future.

Predicting the future with any degree of confidence requires an in-depth knowledge of current technologies, and an understanding of how they might be applied to an ever-changing future landscape. Given the blistering pace of technological advancement, the dynamic could quickly change from predicting the future to merely reporting on it.

It is accepted that exploration is a never-ending game, at least as far as the sciences are concerned. It is completely unrealistic to believe we can find complete answers at the infinitely small end of the spectrum. No matter how deeply we delve into the structure of atomic-particles, there will never be a guarantee that what we observe is not just the behaviour of an amalgamation of smaller components. There is also an area of exploration at the outer-limits where we are permanently denied access. As frustrating as it may be, human endeavour will not discover a way to conquer time and distance before the clock runs out on our species. However, it is certainly an extremely satisfying and exciting pursuit to attempt to discover a model that explains the ever-growing body of data from the outer reaches of the Cosmos.

A dearth of new discoveries in physics is not a problem for applied science. There is a huge backlog of known science that has yet to be fully exploited. Case in point – the thermonuclear processes responsible for the Sun's energy production have been well understood for decades, but duplication of the process here on Earth has remained elusive. Recent experiments to recreate these processes to produce usable energy have offered some encouragement. Within the next few years when the large-scale ITER nuclear-fusion reactor is up and running, it is hoped that the technology will be capable of providing the planet with an abundance of clean energy (ref: **ITER Fusion Reactor**).

Einstein has left us with a very clear message – *Do not become blinkered by focus on the tangible and observable; imagination needs to be free to journey without limitation.* Encouraged by that freedom, it might be appropriate to revisit a scientific theory from the past that has fallen out of favour. In the 1920s, Sir James Jeans first proposed a theory for Steady-State cosmology wherein matter was continuously created (ref: **Steady-State Cosmology**). Some 20 years later, Fred Hoyle and others further developed the theory. They calculated that the current rate of expansion of the Universe would require the creation of only a single atom in the volume of an Empire State Building, each hundred years. Although the idea of something being created from nothing runs foul of laws demanding that energy-matter must be conserved, it would appear that a Big Bang created from nothing takes the breaking of that law to a whole new level. And most in the scientific community today are willing to accept that the Big Bang did start from nothing, or at least nothing we know of. The Steady-State theory was discredited as evidence emerged indicating that our Universe has been predictably ageing from the time of the Big Bang. The Universe itself is expanding, rather than matter expanding outwards to fill the Universe. However, we are no closer to understanding why this expansion occurs. Is the expansion an innate characteristic of space itself, just as mysterious and wondrous as the fundamental forces of nature like gravity and electromagnetism? Or does the expansion result from internal 'pressure' created by existing or newly created matter?

Even though mathematics suggests there could conceivably be additional dimensions within our Universe, there is no guarantee that those dimensions are exclusively associated with our Space-Time framework. They could overlap with other universes, and the whole model could expand to a far more complex system with overlapping universes in parallel dimensions. Even when we restrict our thinking to the observable world, with a single set of dimensions in a single universe, there is still this huge unknown collectively referred to as Dark-Stuff. Any model that does not take this fully into account cannot be considered in the final draft of the 'explanation of everything'. From this point in the history of science, we must accept that many possibilities are open to investigation, thereby creating a fertile field for Philosophers.

The key feature of Dark Matter and Dark Energy is that they are invisible. In this context, visibility is not just a measure of what we can see with naked eyes. It refers to transmission of, or interaction with, electromagnetic radiation. Their existence is postulated to explain gravitational anomalies in the modelling of our visible Universe. So what mechanisms related to the known laws-of-physics could possibly allow something that has mass, and yet does not encroach on the known spectrum of electromagnetic radiation? The current set of accepted laws-of-physics is not able to explain the invisibility of objects with mass, which leaves the field open to theoretical speculation. Could it be possible that some undiscovered particle/field combination might be responsible for visibility? This functionality could be similar to the way Higgs Bosons react with the Higgs Field to produce gravitational effects on objects with mass (ref: **The Higgs Boson**). Atoms are visible because their largest subatomic components are also visible. There is no guarantee that smaller undiscovered particles would be visible before they are able to combine to form the larger subatomic particles that we can detect.

However, Dark Matter/Energy is not the only major cosmic mystery that remains unsolved. In the very early stages when our Universe was forming, current scientific models predict that matter and antimatter should have been created in equal proportions. Antimatter particles have the same mass as their respective particles of matter, but with certain properties reversed. For example, negatively charged electrons have positively charged positrons as their antimatter counterparts. Matter and antimatter particles annihilate when they come together. In one sense, it is a blessing that the amount of matter greatly exceeds the amount of antimatter, or we might not be here to tell the tale. The huge imbalance with matter dominating our visible world is a conundrum. We do not know if the bulk of antimatter is missing, or whether it is simply invisible to us. Quite possibly it could be cloaked within the Dark Matter umbrella. For a time it was postulated that antimatter could have antigravity properties, and thus contribute to the expansion of our Universe. A recent experiment has shown quite conclusively that antimatter particles with mass are attracted to each other (ref: **Antimatter Gravitation**). The search is now on to discover any subtle differences in properties between matter and antimatter that might explain the unexpected imbalance.

The relative abundances between Dark Matter and Dark Energy may well determine the fate of our Universe. Dark Matter is influenced by the contractionary forces of gravity, whereas Dark Energy causes the Universe to expand. The overall expansionary effect on our Universe is quantified with the Cosmological Constant – a positive value implies expansion; a negative value means the Universe is contracting. Recent studies have tantalizingly hinted that the Cosmological Constant may not actually be constant throughout the Universe (ref: **Mapping the Universe**). The jury will be out for millennia before we have an agreed model to predict the end of our Universe. Accepting the inevitability that at some point our Universe will cease to exist, we are faced with a key question that is hopefully rhetorical – Do the Rules of Nature continue to exist beyond the end of our Universe?

Dark matter was postulated to explain observed abnormalities in galactic dynamics. Gravitational forces produced by visible matter alone were not strong enough for some stars to sustain their orbits around a galactic core. If dark matter was evenly distributed throughout the Universe, the overall gravitational effects would balance out and not have an influence on galactic orbits. This observation suggests that dark matter might have followed a similar path as visible matter when forming sparsely populated clumps of matter throughout the Universe. Gravitation has maintained the connection between dark and visible matter, at least on a galactic scale. This does not preclude dark matter from clustering to establish galaxies that are unrelated to visible matter. Because of the association between dark matter and visible matter, it is quite likely that Black Holes contain large amounts of dark matter. In addition to dark matter, recent research has indicated that black holes could possibly be associated with dark energy (ref: **Dark Energy Source**). If we need to rely solely upon astronomical observations to detect dark matter, it could be extremely difficult to distinguish between clustered dark matter and Black Holes assuming there is a difference.

The association between dark matter and visible matter poses some intriguing questions. How deep does this relationship go as regards stars, planets, or even atoms? Does dark matter somehow participate in our daily lives without us ever detecting it? Is the ratio between dark matter and visible matter constant over time and throughout space?

There is a strong possibility that science could advance fast enough before the end of our civilization to develop a model that might explain the influence that dark matter and dark energy have on our observable world. Once that is understood, it could lead to some meaningful projections regarding the future of our Universe. There is a real problem in assuming the rules as currently identified will extend out to the end of our time. We have no way of knowing if every atom of our Universe is destined to head ever outward into oblivion. There may possibly be an undiscovered force that reverses the currently observed process of expansion. When dealing with rules, it is vitally important to appreciate the consequences of there being no changes since time began to the rules of physics, chemistry, and all scientific disciplines. However, it is equally important to consider that rules are scale-dependent, making the results of extrapolation and interpolation from our vantage point somewhat less than reliable.

As noted above, dark entities are very convenient for the development of models to match observations. They allow room to manoeuvre. If there were no fudge-factors included in current models, it would be difficult to marry results from modelling with the empirical evidence. Dark Matter and Dark Energy are postulated to currently exist within our range of observation and measurement. However, there is an equally mysterious cosmic phenomenon that existed outside our range of observation – Cosmic Inflation (ref: **Cosmic Inflation**). Because detailed historical records are not available to us, we must rely almost entirely upon clever modelling and simulation.

As we model-build backwards close to the time of the Big Bang, we arrive at an impenetrable barrier surrounding laws-of-physics that may differ from those of our everyday experience. Most cosmic models include a period of intense inflation, with some scientists going as far as suggesting particles can travel faster than the speed of light during that period. Conditions existed early in the formation of the Universe that will forever be beyond anything that can be simulated on Earth. Empirical verification is essential for theoretical models of physical phenomena to become accepted science, so it is unlikely that the scientific community will ever be able to reach consensus on a model regarding cosmic inflation.

Despite the enormity of the challenge, incredible engineering feats have been accomplished in pursuit of an understanding of the processes involved during the earliest stages of the Universe's formation. It has long been postulated that nucleons break apart at extremely high temperatures to form a quark-gluon plasma, QGP (ref: **Quark-Gluon Plasma**). Experiments over the last decade have achieved the required high temperatures, albeit briefly, and demonstrated that the quark-gluon plasma is a non-viscous liquid rather than the gaseous form that had been anticipated. Most models of cosmic evolution include the quark-gluon plasma as a precursor to the familiar atomic structure of protons, neutrons, and electrons. The transition from plasma to elementary atomic particles during the period of cosmic inflation could be interpreted as a key step in the construction phase of the Universe. Confirmation that the QGP is real raises some intriguing questions. At a time when it presumably permeated the entire Universe, where was Dark Matter? Did Dark Matter coexist independently, or was it somehow produced as a by-product when the plasma cooled to create visible atomic matter? Experiments with high energy particle accelerators may be able to answer that question in the coming decades.

Although temperatures in the experiment noted above are breathtaking by Earth-bound standards, we have no way of knowing how high the temperatures might have been at the birth of the Universe. The temperature scale is open-ended. The lower end is clearly identifiable and associated with a lack of heat-energy. The unimaginable density of heat-energy, or any other kind of energy, at the start of the Universe renders the concept of temperature rather meaningless in that context. Even though it seems increasingly likely that the quark-gluon plasma phase existed at some stage during cosmic inflation, the processes that preceded it may never be revealed.

To build a credible model for the cosmic inflation bubble and the processes within, a good starting point is to assume that the established laws-of-physics apply until they obviously don't. The duration of cosmic inflation is estimated to have been an unimaginably short time – around 10^{-36} seconds (ref: **Cosmic Inflation**). Reasoning back from the Universe we know today gives an approximate size of the Universe at the end of this inflationary period. As noted above, scientists have taken these figures and calculated that particles must have been able to travel faster than the speed of light during inflation. Einstein's theories of Relativity have withstood the test of time (literally). If the laws-of-physics from everyday experience apply equally during cosmic inflation, time-dilation within the near-infinite gravitational field close to the Big Bang would result in time virtually standing still for part of the inflationary period. The estimate of the inflationary period could be quite deceptive. From a human perspective, calculations of split-second duration could be quite accurate. However, to an observer inside the cosmic inflation envelope, it might have occurred over billions of years.

Consider a time very early after the birth of the Universe where atoms did not exist, and there was nothing available for the construction of the elements currently visible throughout the cosmos. What is truly remarkable is the fact that the rules dictating the behaviour of atoms were in place long before atoms ever existed. Similarly, rules governing the behaviour of everything must have existed all the way back to the earliest moment of the Universe's creation. It would seem that the rules are intimately tied to the Space of our Universe and exist outside any physical manifestation of matter.

The Big Bang could be considered to be an event where the rules for behaviour of all things were created afresh. As the Universe expands, it is a big stretch to imagine how the rules might be able to vary over time. If they did, it would mean a different set of rules applied at the start of our Universe to those that apply at the end. Examining the available data covering almost 13.8 billion years suggests there is no variation in the rules. But again, we need to be convinced that this is not too short a time-frame to make a judgement. This then presents a challenge for science to devise some technique to identify any subtle changes in the established laws-of-physics since our time began.

There is an analogy that might be appropriate for modelling the initial moment of the Big Bang. In a kitchen experiment where a cup of water is superheated in a microwave oven, the water might sit at over 100 °C without displaying the familiar bubbling and change-of-state to gas. At the slightest disruption to this unstable equilibrium, the cup of water will erupt. Imagine that the composition of the early Universe is equivalent to the water in this unstable state. The Big Bang could have been triggered by a random perturbation that produced a violent reaction in this unstable medium, and we are still witnessing the after-effects that flowed from that chaotic beginning.

Across 'recorded history', most of the identifiable ages of mankind have been associated with some major scientific breakthrough. A clear understanding of electromagnetism heralded in the age of electro-mechanics and radio transmission. During the modern era, an intimate knowledge of semi-conduction produced the transistor, which gave birth to the electronic age with computers and a myriad of display devices. This perhaps raises some interesting questions for academic debate – Which 'age' comes next? What new scientific discovery could be so dramatic that it ushers in a new age for humankind?

RULES FOR ETERNITY

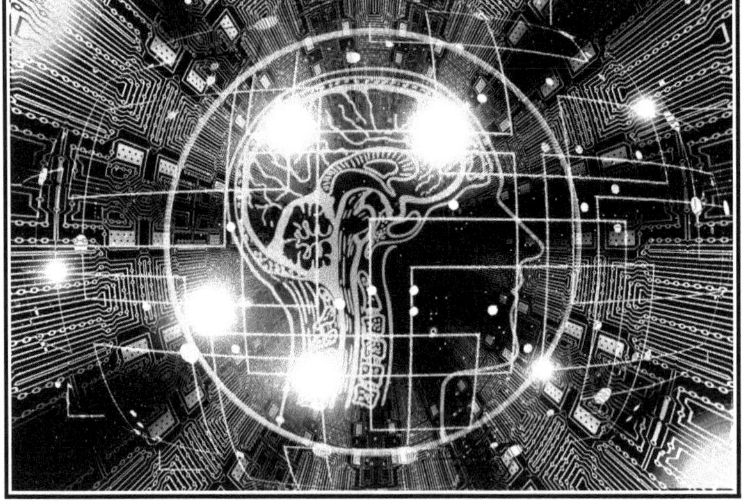

9. APPLYING SCIENCE: THE CHALLENGES

The previous chapter covered some of the big issues of concern to physicists at the leading edge of their profession. Applied scientists and engineers are more closely focused on the practical benefits that can be derived from the underlying science. Scientific endeavour is split between Engineering and Reverse Engineering. For particle-physics and molecular biology in particular, deep probing in the search for fundamental building blocks is an exercise in reverse engineering. Investigating new ways of assembling these building blocks has led to much of the innovation evident today.

An interesting area for the application of science is the development of robots that mimic or even enhance human functionality. The goal in building a human-like robot is not necessarily to produce something that functions exactly the same as a human in every respect. And neither is it purely for the appealing novelty of the contraptions. For example, developing and experimenting with these devices is certain to produce tangible results to aid humans who might benefit from prosthetic assistance.

Engineering of humanoid robots embraces many disciplines...

- Investigation of suitable strong, lightweight materials
- Development of lightweight, high-capacity power pack
- Coding of computer algorithms gleaned from the study of animal behaviour
- Research into wearable materials and devices for eventual human attachment

Any machine designed in humanoid form to solve a specific task is likely to be clumsy in the extreme. Humans have been generically designed for a vast array of applications, although performance may not be optimal for many of those applications on their own.

In general, robots are designed to perform specific tasks that normally involve human activity. Robotic devices have been produced for a huge number of applications, but most of these devices have one restriction in common – they need to be tethered to a power source. Mechanical robots differ in some major aspects from their biological counterparts. One of the main differences relates to power source. A biological source of power has the advantage of flexibility – fuel is usually readily available in many different forms. Unless a robot is tethered, it must either rely on limited solar-power or stay within reach of a suitable power outlet. Another distinguishing feature is that biological mechanisms have a certain degree of capacity for self repair. Science Fiction might have us believe that we are close to a mechanical solution to that problem, but the reality is that such an ability will not be available until well into the future. In the meantime, we can alleviate the possibility of failure in robots by including ample redundancy.

So what characteristics of animals have been identified as important in the pursuit of technology for autonomous robots? Most animals are blessed with a command-centre, typically recognizable as a head. As with control-towers at airports, it makes good sense to have that command-centre, including its various sensory devices, located as high as possible. This allows the best way of assessing factors that play a part in an animal's survival and well-being. The primary functions of an animal's body are to provide power and to enable mobility. One thing the science fiction guys often got wrong was to show the head of a humanoid robot continuing to function once it has been separated from its power-source. Depending on the degree of semi-automation built into the robot's body, a headless robot might continue operating for some time, although in a less predictable manner. In sophisticated mobile life-forms, the head is so jam-packed with complex componentry that there is no room for a backup power supply. Normal operation of the head would cease instantly when disconnected from its power source.

The human brain has amazing processing ability, incredible memory capacity, and communicates with all sections of the body through a vast network of nerves. Despite these attributes, there are limits. In the interests of efficient design, some body components have a degree of self-autonomy through a process referred to as 'muscle memory'. In less sophisticated mobile animals with smaller brains, bodies regularly rely on semi-autonomous components to reduce the workload on the brain. As an example, consider a chicken whose head has just been chopped off. The body might appear to run around for a short time as if all is fine, albeit with noticeably impaired coordination. To facilitate walking for a headless chicken, leg movements must be more complex than just twitching backwards and forwards. To maintain balance, each leg must be aware of what the other leg is doing. This raises an intriguing question – in the absence of a brain to coordinate movement, how do the legs communicate? Definitely something worthy of future research, as long as animal-rights groups do not object to the proposed experimental procedures. The more we understand about the biological processes involved, the better we can design prosthetic replacements for defective human componentry.

When designing mechanical robots for a particular spatial terrain, the result is likely to be quite specific, and not nearly as flexible as the implementation for animals. There is one thing in common between animal and mechanical design. For both, a careful evaluation of the environment is a necessary precursor to the design. To determine what life-form is best suited to a particular environment, there is a vast portfolio of possibilities to choose from. Engineering is in its infancy when it comes to developing a complete set of building blocks for all-purpose robots.

Silicon-based binary computers simulate the processes that are part of our human experience. They take a given set of inputs, process those inputs according to some inbuilt algorithms, and produce outputs that can be interpreted by humans. The algorithms involved mimic the functionality of the rules of nature in that they prescribe the range of outcomes permissible with a given set of inputs. Recent research into Quantum Computers offers the potential to increase computing power by many orders of magnitude (ref: **Quantum Computing**). Binary computers are relatively predictable in that the same inputs processed by the same algorithms will produce essentially the same results. Although quantum computing promises exciting possibilities, the price to pay may be that the results are not quite as reproducible as they might be with their binary counterparts.

Many computer applications do not require extremely accurate results, and sometimes benefit from a level of chance variation. Typical of these applications might be when computers are used in modelling weather-patterns, or to simulate the evolution of our Universe. Both types of computers have parallels in the natural world. If we experiment with chemical reactions under very controlled conditions, the expectation is that the same experiment can be conducted many times and produce exactly consistent results. If we create a precise set of conditions where new life is born, the results are likely to be quite different every time, given the limitations on our ability to ensure that all conditions are identical.

When coding computer algorithms, we are defining the rules that dictate computer behaviour. For binary computers, the instructions are usually quite explicit. A new technique is leading towards Artificial Intelligence (AI), where computer programmers seek to emulate the learning processes involved in organic brains. In broad terms, previous results are fed back to modify the original instructions written to define behaviour, or change the computing environment in some way. It is important to note that in order for code to self-optimize and 'learn', instructions to enable these features must have been written into the original code, most likely by human programmers. Any danger posed by this artificial intelligence becoming REAL intelligence, is a long way off, and is often sensationalized and over-dramatised in the media.

Much has been written in recent times about the threat of computers becoming more clever than humans. This has tended to mask the real issue where humans are becoming more like machines. For the majority of Earth's inhabitants, the Internet has provided easy access to a vast amount of information. What computers have done best in the past has related to storage, collation, and processing of data. There is currently a paradigm shift underway that allows computers to insert a layer to interpret and format information before making it available for human consumption. Problem is, this interpretation layer is easily corruptible, whether that corruption is intentional or not. Future success of artificial intelligence is entirely dependent on its ability to distinguish between information and misinformation. This is an ever-growing challenge, and it may turn out to spoil the party for AI. Any threat posed by artificial intelligence is likely due to the collective decline in human ability to assess the value of information they are presented with. Quality scientific evidence and assessment is all too easily given equal status with social media hearsay, leading to an inevitable degrading of the message.

There is a negative consequence of humankind being immersed in a cradle of managed information – complacency. Back in the days when one needed to devote serious effort to access quality information, there was a corresponding effort given to logical assessment. The sad reality is that clear-thinking has taken a back seat, and many people today are far too willing to accept without question what the media and Internet present to them. Lemming-like, we often seem to be joining the mindless Mexican Wave, and it is difficult to identify anything that might reverse this trend.

In an earlier chapter, we posed the question as to what the next Age of Man might be. Perhaps the next stage might be characterized by extended integration between humans and electromechanical devices. For some people this might suggest a doomsday scenario.

However, for most technological advances there has been both an upside and a downside.

From their emergence, humans and similar bipedal primates have primarily relied on their hands for tactile interaction with the environment. Hands are incredible mechanisms capable of precision movement. Part of the reason for this high precision is that hands provide essential sensory feedback to the brain by augmenting the other senses. Over time, tools have been developed to make full use of that precision, and extend the capability. Joysticks, the computer mouse, and touch displays are now standard fare in our daily lives. Current state-of-the-art has reached the stage where surgeons can remotely perform the most delicate of operations on patients. Technology has also provided a system capable of controlling devices using the human voice. But the story about humans interfacing with machines does not end there.

At first glance it might seem that mind-control could provide the ultimate solution for connecting human with machine. There are two major issues that may prevent mind-control being practical enough for mainstream adoption. Firstly, the engineering challenges are formidable. Signals from the brain are extremely weak and hard to detect outside the head, and those signals are competing with a background of extraneous brain noise. The second issue confronting mind-control is the brain itself. Even a well-disciplined brain hosts a cacophony of wild activity that could thwart any attempt to decipher the intent of a particular signal. However, all is not lost. Although it might be virtually impossible to detect individual signals related to individual thoughts, it should be feasible to train the brain to think major thoughts capable of producing recognizable signal-patterns. If a set of unique brain-signal patterns can be correlated with specific thoughts, it would be a simple matter to configure an external device to respond in a certain way to each signal. In this context, a "major thought" would result from a subject's concentrated effort to consolidate thoughts in some particular way, thereby producing strong signals that could be distinct and easily detected.

Over the last decade or so, much research has been done using external sensors to discover a link between thought patterns and measurable brain signals. However, the results have been patchy, and the engineering challenges remain. Trials have recently begun to insert brain implants in patients with severe physical disabilities. It is hoped that this technique can more clearly define the relationship between thoughts and brain signals. Many implantable devices to monitor or enhance body functions are already available. The new research differs in that it attempts to extract information that is directly under control of the patient, and is generated without any physical movement. Implants might seem to be quite invasive, but for patients most likely to benefit from the procedure, it is a small price to pay. No matter how accurately brain signals can be captured and interpreted, mind control is never likely to match the precision possible with a healthy pair of hands. Benefits of experimenting with implants in the interpretation of brain signals go well beyond the development of clever man-machine interfaces. The human brain is the most complex organ we are ever likely to encounter. Discovering the fundamental biological rules of nature responsible for brain functionality will occupy scientists until our time on this planet is up. Analysis of data from future brain-implant trials could deliver a valuable contribution during this journey.

A timely postscript to this chapter...

As the human knowledge-base grows, we are facing a serious limitation due to the mismatch between the rate of evolution and the rate of growth of that knowledge-base. No reference to the knowledge-base would in modern times be complete without mentioning the Internet. If the promotion on the packet is to be believed, the rising tide of available information should make all participants smarter. The problem here is that the wealth of valuable information is equally offset by a huge amount of useless misinformation. Pursuit of any scientific discipline requires a significant percentage of one's working life to gain background sufficient to make a contribution to the knowledge-base. We are stuck with more or less the same brain capacity we have had for the last few thousand years. Evolution, combined with the significant time between human generations, is not likely to offer a short-term solution. Could this result in a slowing of the exponential growth in the knowledge-base that has been building over the last few decades? Mentioned in an earlier chapter was the benefit to society from conformity. One hope for the human race is that we all start working together, and that requires communication. The Internet certainly provides the mechanism for that communication, but human traits may hinder cooperation on a global scale.

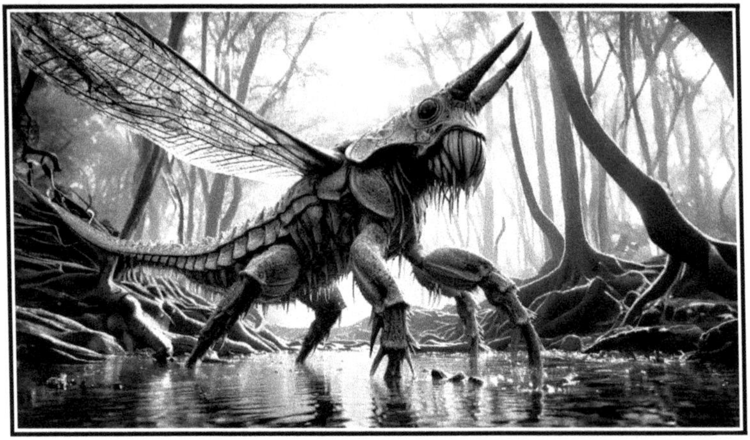

10. WHAT IF THE RULES WERE DIFFERENT

No Rule in the Natural World is without purpose, even if that purpose might not be apparent from a human perspective.

Life as we know it here on Earth has resulted from two very distinct stages of development that occurred on very different timelines. Nature's biological rules were nowhere to be seen in the early Universe before an environmental framework suitable for emergence of life had been established. As a precursor to life, complex atomic and molecular structures were produced by maybe a billion years or more of cosmic evolution. Next step was to arrange delivery of the chemical soup to a rocky planet that was capable of incubating life. To kickstart life on this planet, there is a long list of tight-tolerance specifications that must be met. Given that the formation of stars and planets is a process with a myriad of possible scenarios, the likelihood of finding the perfect planet to sustain intelligent life is extremely remote. And yet, here we are.

Humans have existed on Earth for a very long time, at least in terms of the way humans perceive time. That has allowed humans to become so familiar with their environment that it is easy to underappreciate the precise confluence of factors that led to our existence. The slightest modification to the rules of nature or a subtly different environmental framework would have created an unrecognizable landscape of life, or no life at all. In the interests of highlighting the incredible good fortune humans had in establishing a base on Earth, it might be instructive to contemplate the ramifications if any of the fundamental rules were slightly different.

According to current scientific thinking, there are four fundamental forces of nature – Gravity, Electromagnetic Force, Strong Nuclear Force, and Weak Nuclear Force. The most significant of these on a cosmic scale is gravity. It is responsible for bringing together primordial components and compacting them to establish stars. Stars make a twofold contribution along the journey towards intelligent life. Firstly, they are capable of hosting planets that might potentially be able to sustain life. In addition, stars are nuclear-fusion reactors that produce some of the heavier elements that are essential for carbon-based lifeforms. Any subtle decrease in the force due to gravity could prevent particles from coalescing to form stars, and end all hopes that this universe would contain intelligent life. A small increase in the force due to gravity might not be quite so devastating. Localized effects on lifeforms due to a planet's gravitation could be compensated by specifying that the 'ideal' planet size for life is a little smaller.

In the discussion that follows, "life" is generally to be taken as a reference to intelligent lifeforms, i.e. advanced animal species with a high degree of cognitive ability rather than low-level microorganisms. Planets with the potential to support life tend to settle into a stable orbit in the Goldilocks Zone, just far enough away from the parent star to receive precisely the right dose of light and radiant energy. However, existence in close proximity to a giant thermonuclear reactor does come with consequences. During most of the productive life of a star, it emits a steady stream of charged particles known as the solar wind. Any planet that does not have a defence mechanism to shield from this solar wind is extremely unlikely to support intelligent life as we know it.

Solar winds are detrimental to life in two ways – they have the potential to destroy a planet's atmosphere, and also contain charged particles that pose a direct threat to cellular life. The atmosphere is an essential part of the ecosystem for life. It surrounds a planet with a delicate mix of chemical elements that must be carefully managed and maintained long enough for life to evolve. Intense ultraviolet radiation from the parent star is detrimental to many surface-dwelling animal species. On Earth, for example, serious cancers can result from unrestricted exposure to the Sun's rays. A healthy atmosphere aids the development of life by absorbing much of the harmful radiation before it reaches the planet's surface. Solar winds can be sufficiently strong to strip away the atmosphere, including the all-important component, water vapour. The atmosphere is capable of filtering much of the undesirable radiation, but is less effective in providing protection against particle bombardment. Charged particles radiating from the parent star have a negative impact on any life that may have previously gained a foothold. In the absence of a shield to deflect a large proportion of these particles, they could wreak havoc on cellular life and produce a devastating disruption in the cycles of evolution.

The protective shield between star and life-supporting planet comes courtesy of another fundamental force of nature – electromagnetism. Again, using Earth as an example, circulating liquid metal in the core produces an electrical current that creates a magnetic field surrounding the planet. Because particles from the Sun are charged, Earth's magnetic field is able to deflect them to a large degree. This planet-wide effect is the largest known demonstration of electromagnetic force in action. If it is the only consequential application of electromagnetism on a cosmic scale, a slight increase or decrease in strength of the force could be tolerated without having a serious impact on the Universe or its ability to foster life.

Pushing speculation to the limit, it might be conceivable for other universes to exist with slightly altered parameters for gravitational and electromagnetic forces, and yet still appear remarkably similar to our own Universe. What about the other fundamental forces of nature – the Strong Nuclear Force and the Weak Nuclear Force? The Strong Nuclear Force is responsible for keeping nucleons tightly bound in the atomic core. Without this force, it is hard to imagine how atoms might exist in an alternative reality. Chemical composition and interactivity are controlled by the Weak Nuclear Force. These forces combine to define every aspect of observable matter in our Universe. The strengths of the forces binding atomic and subatomic particles together, must be extremely precise. If they were not, atoms would either collapse or fly apart, and nothing in our existence would be possible. There is zero tolerance for variability. This inflexibility should put paid to any notion that the rules of nature might have been tweaked and fine-tuned prior to application in our Universe.

On a galactic scale, stars exhibit a type of lifecycle where remnants of dying stars are ejected for recycling in the birth of new stars. During the life of a star, there is an opportunity to form planets conducive to life. These self-contained planetary ecosystems provide the laboratory for biological experimentation that could lead to a vast variety of species, all having their own unique lifecycles. However, the biological lifecycle and cosmic lifecycle are very different. Inheritance, generational life-cycle, and natural selection, combine to produce the truly wondrous phenomena of biological optimization. Galaxies, stars, and all non-biological entities in our Universe follow the pattern of behaviour ascribed to them at the beginning of time. In other words, no optimization is possible.

Thus far in this chapter we have considered consequences of varying some of the high-level rules that govern existence. There is a relatively independent subclass of rules devoted entirely to organic life – the biological rules of nature. Suggestions for ways to modify life-controlling rules could result in a long list. It is important to make the distinction between possible modification of the rules, and medical experiments involving observation of behaviour that results from application of the existing rules. In order to constrain the discussion, the approach taken below is to examine some aspects of the human experience as supplied by nature, and compare them with solutions that might be derived from human inventiveness.

In contrast to the constancy and immutability of Nature's rules, planetary environmental conditions are in a continual state of flux. As a result of this, any life supported by that environment must adjust accordingly. For complex life to form, the evolutionary process must coordinate with the rate of change of the environment. If evolutionary development is too slow to keep pace with environmental changes, intelligent life may have difficulty establishing a foothold. This delicate dance between evolution and the environment has served humans well up until now. Factors related to industrialization have precipitated a rate of environmental change that the human evolutionary process had not anticipated. Unless a concerted effort is made to subdue the environmental change-rate, future generations will struggle to cope with the mismatch between environment and evolution.

Out of context, some rules might seem perplexing, and their purpose questionable. Very clever reverse-engineering is often required to determine the exact operation and purpose of a particular rule. As we provoke the rules into action under various circumstances, we can infer the rules from the consequences of that provocation. Humans are never likely to discover the complete set of rules with their accompanying sets of subtle qualifiers. This journey of discovery moves ever closer to the inevitable conclusion that everything in the natural world has a purpose. The laws of physics and chemistry work harmoniously to support physical existence. Biological mechanisms operate to create and optimize life wherever conditions are suitable. When life does become established, it either makes a contribution to benefit the environment, or fits neatly into an appropriate slot in the food-chain. However, there seems to be one glaring exception to this concept that everything has a purpose. Humans, being animals at the top of the food chain, vastly over-consume valuable resources and do nothing to help the environment, appear to defy the requirement for purpose.

Darwin's Theory-of-Evolution, with its embrace of 'natural selection', has been thoroughly researched and documented. However, there are two additional factors that have a vital influence on the possible outcomes for life – rate of mutation, and the amount of time between generations. For humans, a generation lasts approximately 25 years. Compared to most other life forms, this is quite long. A serious challenge to the long term survival of humans comes from the fact that while bacteria and viruses might have the same rates of mutation, the timespan between generations might be counted in hours rather than years. If the situation arises where these tiny organisms chart a course that is detrimental to our species, the odds very much favour a win for the little guys. There is another downside to the slow rate of evolution in humankind. Growth of knowledge is now proceeding at such an incredible rate, that there is no way the capability of the average brain can evolve fast enough to accommodate the rapid increase. How our species might adapt to this challenge is yet to be determined.

The theory behind Rules-for-Eternity requires a complete break with the notion that the blueprint for any particular lifeform was on the drawing-board well in advance of its existence in the physical world. There are many examples of curious phenomena that might lead to the conclusion that some particular lifeforms have been preplanned. Take for instance the configuration of the human body. Externally it has bilateral symmetry in a vertical plane, a feature of all mobile creatures. 'Balance' is a fundamental requirement for optimal mobility techniques, and no human design team has ever managed to improve on the way it has been implemented across the board in Nature. In fact, humans closely followed the bilateral-symmetry model right from the invention of the wheel and subsequent wheeled devices. No question – bilateral symmetry is an optimal design feature.

Let's examine the process of optimization by supposing we might be tempted to change some design rules in the hope of achieving a better result than that delivered by Nature. Consider the situation for highly mobile, intelligent creatures who walk upright – should they have one leg, two legs, or even more? As a way to illustrate the challenge, imagine a group of engineers tasked to build a humanoid from scratch. A logical starting point would be to give it three legs, just from the perspective of stability. For a two-legged model, quite complex control-mechanisms would be required to produce any practical level of stability. In a real human, these controls are already in place for a multitude of reasons. The choice between increased complexity or additional hardware (more legs), would not be an obvious one. The optimal design would require extensive knowledge and masses of empirical data about the environment in which the humanoid is to operate. The current design for human beings does appear to be optimal, taking into account all situations where humans might apply themselves well into the future.

The answer to the leg-count question would seem to have been available before the question was even asked. Could we assume that human-like creatures are destined to have two legs because there is a rule of nature to cover this? In answer, probably not. Animal and plant species have many respective features in common. For example, they are both composed of cells, and use DNA structures to convey instructions for evolution and development. Is this convergence of biological design the result of extensive trial-and-error experimentation that produced the ultimate optimization? If a certain design feature is obviously optimal, it would be logical to expect the design to be replicated wherever appropriate. Here we face one of the most serious difficulties in defining what actually constitutes a rule. Some functions in nature could be classified as rules in their own right, while in effect they are purely the result of optimization using previously identified rules. There is another interesting implication related to the optimization – given some rules can be both optimized and immutable suggests that any optimization should have occurred outside the framework of our current Universe.

Suppose we had the power to change a few of the biological rules of nature that have a direct influence on our existence. Imagine for a moment that the average lifespan for humans was 200 years. Would individuals in the late stages of their life-cycle be capable of contributing to the benefit of mankind, beyond that of someone say in their 50s or 60s in today's world? If we subscribe to the tenet that all the rules have already been optimized, then a dramatic increase in average lifespan would not benefit humankind as a whole. An extended lifespan is also going to seriously grow the population, and put further demands on planet Earth's scarce resources. A recent report highlights that current consumption of resources is running way ahead of anything our planet is capable of delivering (ref: **Unsustainable Consumption**). Further suppose that science and technology can provide a short-term solution. Eventually a time will come when a balance between population size and resource consumption cannot be sustained. The point of this illustration is to show that the rules of nature will always have the last word. Even if we populated a planet which had 10 times the usable resources available on Earth, there would still be some point where population growth would need to be restrained.

Would humans be more intelligent if they were given bigger brains? Perhaps they might, but what negative impact might a physically larger brain have on the overall design? Brain function has been compartmentalized, and the componentry condensed to produce what is arguably the most efficient biological design possible. Assuming there is no opportunity for optimization, a bigger brain would take up more space, and could lead to the human animal becoming top-heavy. Top-heaviness would not be an issue if mobility and agility were not of primary importance. It is theoretically possible to grow a giant brain with plant-like immobility, which then raises the question as to the purpose of intelligence if there is no mechanism to share that intelligence. All biological design has been derived from a holistic approach.

We are just beginning to understand how finely tuned humans are to conditions on planet Earth. Analysis of the health of astronauts who have returned from extended stays in space reveals how circulatory systems have been optimized for vertically-orientated humans who experience a constant gravitational force of 1 G. There is a multitude of knock-on effects resulting from any disruption to the steady distribution of blood throughout the body (ref: **Neurology of Space Flight**). An interesting design exercise would be to develop a computer model to predict what humans might be like after evolving for a few thousand generations in microgravity.

Operational areas within the human brain are not symmetrically arranged, so it is not particularly surprising that functionality, like coordination for example, might not be identical for both sides of the body. All brains belonging to specific species of advanced lifeforms exhibit the same pattern of asymmetry. Left-handedness could come about purely by chance, but the location of human organs conforms to a very specific layout. Organs always end up in the same place for each particular individual species. How this came about remains one of the major unsolved mysteries of biology.

Externally, humans are bisymmetrical, but not so for the internal organs? Many organs within the human body are, of necessity, laid out asymmetrically. That in itself might not be remarkable, but what is absolutely astonishing is that the asymmetry is consistent across the entire species. Darwin's Theory of Evolution involves a trial-and-error process that gives bias to the option that has the best chance of survival. And here's the conundrum – if a human had internal organs laid out as an exact mirror image of the norm, would they not have an equal chance in the competition for natural selection? How come there is no evidence that this option was ever considered? Admittedly, the decisions for organ layout were probably made way back somewhere in our animal ancestry, but at some point a choice must have been made to go with either one option or its mirror layout. Why could the two options not have evolved side-by-side since there appears to be no advantage of one over the other? This suggests that all humans relate back to a single ancestor, and that is where the belief in a blueprint for animal design gains credibility.

Perhaps the hidden part of this picture could be that there is some obscure reason as to why the current layout of internal human organs is actually better than its mirrored counterpart, but we just haven't discovered it yet. Rare mutations cause some humans to be born with mirror image layouts for their internal organs (ref: **Situs Inversus Totalis**). If there is no disadvantage with this configuration, one might expect the two to evolve in parallel as genes disperse into future generations. There does not appear to be any evidence that this has occurred.

In addition to Pressure and Temperature, there are less obvious contributing factors that are essential if an environment is to nurture and sustain life. A subtle but very important influence comes from rotation, whether it be controlling galactic arms during the formation of stars, the orbit of planets around a star, or the rotation of planets themselves. Two of these effects have a direct impact on life on Earth. Planets depend very much on rotation around their own axis to provide stability. Without this stability, there would be no predictability about the temperature ranges possible in various zones. The concept of North and South Pole would be lost, and temperatures in any part of the world would be at the mercy of the randomly unpredictable orientation relative to the Sun. All life on our planet is intimately tied to predictable cycles, and evolution relies heavily on the future repeating the environmental patterns of the past.

Just as important as stability is the dependable cycle of night and day due to Earth's rotation. There is a very identifiable pattern in virtually all animals with an average life span longer than a few days – they are all geared to some form of rest and recuperation on a daily basis. It is probably not valid to draw any conclusions about what role the actual length of a day might play in determining possible options for life. In fact, it is highly likely that the vast majority of observed forms of life are as they are, precisely because they have evolved under the influence of regular cycles dictated by the environment. Virtually all species have developed their own unique rhythm in response to the prevailing environmental conditions. One possible exception is life deep in the ocean, well shielded from cyclical processes.

So considering all the factors that have to be just right for the formation of life on our home planet, is it valid to assume that exactly the same conditions are necessary for intelligent life to develop elsewhere in the depths of the Universe? Modern science and discovery teach us just how tenacious life is. There is a huge range of possible environmental parameters, and the rules for life are flexible enough to sometimes produce results in the most unexpected ways, even in extremely hostile situations (ref: ***Extremophiles***). Nature's biological rules are so accommodating that it would be extremely difficult to imagine how they might be modified to increase the chances for life to exist.

To everything there is a season, and a time to every purpose under heaven:

> *a time to be born and a time to die*
> *a time to plant and a time to uproot*
> *a time to kill and a time to heal*
> *a time to tear down and a time to build*
> *a time to weep and a time to laugh*
> *a time to mourn and a time to dance*
> *a time to scatter stones and a time to gather them*
> *a time to embrace and a time to refrain from embracing*
> *a time to search and a time to give up*
> *a time to keep and a time to throw away*
> *a time to tear and a time to mend*
> *a time to be silent and a time to speak*
> *a time to love and a time to hate*
> *a time for war and a time for peace.*

Bible: Book of Ecclesiastes 3:1-8

Byrd's Song Lyrics: To Everything There is a Season

RULES FOR ETERNITY

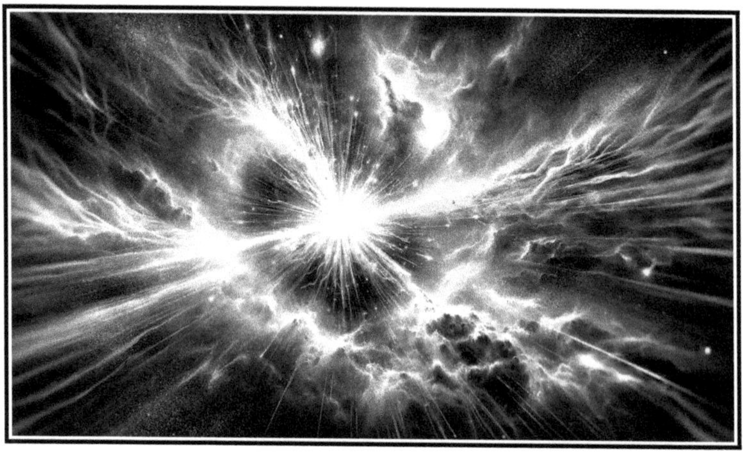

11. BIG BANG AND BEFORE

Humans will never be able to observe cosmic events in real-time. We are speculators rather than spectators.

Today, it is near impossible to argue against the accepted scientific model in which the space and time of our Universe began with the Big Bang. Any models attempting to explain the environment and processes prior to that event will fail to find empirical validation. However, that does not discourage examination of the possibilities. Even outside our Universe, the seemingly unavoidable assumption would be that all processes are governed by a set of rules. So what unique rules might apply to a higher-level system that could contain universes similar to our own? Let us speculate about the composition of this uber-universe or "Uberverse", and also about the possible set of associated rules. Rules only become known to us because of their influence on the behaviour of tangibles within our range of detection. It is quite conceivable that the rules to create universes are ever-present, but in the absence of special circumstances required to activate those rules, they will remain dormant and thus invisible to us. At the very least, the Uberverse must provide an environment suitable for the operation of rules that enable the formation of universes.

In its most basic form, the Uberverse should have spatial dimensions, and some distribution of matter-energy throughout that framework. Spatially, it could be either bounded (linearly or spherically constrained), or else might extend without bounds. To imagine a Uberverse that did not involve spatial dimensions would be extremely difficult. A lack of separation in space would imply all matter-energy congregates at the same place. This would suggest that our Universe could be the only one in existence at this particular time, and not sit well with the concept that there could be a multitude of parallel universes. If we assume in our model of the Uberverse that it contains some form of matter-energy, then that matter-energy would need to be distributed. And distribution implies the existence of space to be distributed within.

One of the fundamental laws of physics applying within our Universe states that mass-energy in a closed system can be neither created nor destroyed. If this rule applies across the transition between Universe and Uberverse, it could offer some valuable insight into environmental conditions existing moments before the Big Bang, at least in terms of the concentration of matter. It offers no clues regarding any change of state that might have occurred during the transition. Before the Big Bang 'ignited', the Uberverse could have been in a state of relative calm, or it might have closely resembled the intensely chaotic plasma of subatomic particles believed to exist at the birth of our Universe.

In the absence of any clear definition or understanding about the main constituent of the Uberverse, we refer to it using the intentionally vague term – "matter-energy".

Theoretical models of the Universe became a lot more complicated when it was discovered that the actual space within our Universe is expanding. We can no longer rely on the earlier simple model where all matter contained within our Universe is moving outwards from a central point into some larger, undefined empty space. For both the earlier model and the currently accepted model of our Universe, the burning question remains – what does the Universe expand into? The real problem is the way we have been conditioned to think of spatial dimensions as basically linear and reliably measurable. Let's pose an intriguing question, which, although it invites highly-speculative answers, could deliver useful suggestions about how to find some rational connections with known science: Does our Universe take up any space in the framework that surrounds it?

Internally, we estimate the volume of space within our Universe to be somewhere between almost zero, and countless quad-zillions of cubic light-years. In other words, it is really unquantifiable. What appears to be constant for our Universe, and possibly even measurable, is the total amount of matter-energy contained within. Externally, our Universe may not occupy space as such, but could be considered as a containment of a specific quantity of matter-energy. This conclusion might actually be quite comforting as it avoids the depressing outlook that our lifeless Universe could eventually expand to fill the entire Uberverse, thus permanently terminating any future opportunity for existence.

The Uberverse could expand like our Universe, it could contract, or it may not change in size at all. If co-ordinate space has any meaning at all in the Uberverse, as weird as it may seem, the most plausible model for the Uberverse might be based on a non-expanding framework that extends to infinity. Theoretically, anything that stretches to infinity cannot expand further. Similarly, there is a serious conceptual problem for anything that reaches to infinity and then starts contracting. The corollary here is that anything not stretching to infinity must be bounded. There is another argument against the notion of an ever expanding Uberverse – decreasing density. If the Big Bang eruption was triggered by an intense accumulation of matter–energy, such accumulation would be far less likely in an expanding framework. These ideas might seem quite strange, and unrelated in any way to human experience. However, the bond between strangeness and impossibility was broken by the discovery that the space in our Universe is expanding.

Where Infinity is involved, its natural bedfellow is Eternity in the time dimension. Humans are very much preoccupied with the quantification of time. In a time-frame that stretches to eternity, the actual rate at which time passes has little significance – it really does not matter if time passes slowly or quickly. The period of time between the start and end of our Universe, is hardly distinguishable from the time it takes to boil an egg. The Uberverse could be expected to operate without urgency, and with no predetermined schedule that must be followed.

In support of the idea that the rate at which time passes is of little consequence in the Uberverse, consider conditions at the earliest moments after the Big Bang. As mentioned in a previous chapter, the passage of time during the initial formation of the Universe could have been slower than a snail swimming through a block of ice. The same laws-of-physics relating gravitational forces and time-dilation may also have been in effect in the moments just prior to the Big Bang event. However, there is a major problem with this assumption. Space, Time, and Gravity, are intimately intertwined and bonded with the Universe, and are absolutely essential to the existence of life. Space and time may both have equivalents in the Uberverse, but the role or necessity for gravity is far from obvious.

In the realms of unverifiable astrophysics, anyone can propose the most outlandish models to explain the unknown (including this author). At the end-of-the-day, models likely to gain acceptance are those which most closely align with established scientific understanding. Without at least a tenuous tie to known science, theoretical physics is indistinguishable from pure science fiction. With that in mind, it would seem logical to start building a model for the distribution of matter-energy in the Uberverse based on observations within our own Universe. Fundamental to the function of the Universe is Space, with its curious ability to warp and produce gravitational attraction. Gravitational dynamics would not be relevant in the Uberverse if it was completely homogeneous, with no variation in the concentration of matter-energy distributed throughout.

With an endless canvas to paint upon, it would seem an unnecessary complication to have the Uberverse contained or constrained in some way. To align best with the human view of reality, containment of some kind might seem logical, but that inevitably leads to the question of what exists outside the container. Returning to a recurrent theme in this book, it would seem like an absurd waste of effort to create just a single universe with an expiration date. Philosophizing once again, the most logical model for a Uberverse would be one with no time or spatial constraints, and an unlimited number of recycling processes capable of spawning universes like our own.

Every aspect of experience in a human life can be related to a beginning and an end. Many people consider the creation of our Universe as the ultimate beginning. Logic dictates that, at some point, the Universe will end. The end of our Universe cannot be empirically probed, so all theoretical models that attempt to describe the process must forever remain the subject of conjecture. From a philosophical standpoint, we should have a better chance of discovering details of the Universe's origin from historical records, than we would have in predicting an outcome for which there is no available precedent.

Without question, our Universe had a beginning, and there is every expectation of an ending. In discussions regarding the end of the Universe, things get really interesting when questions are raised about what it is that actually ends. Presumably all the matter-energy within the Universe is preserved, so it might appear that nothing actually ended – it's just the container keeps getting bigger and bigger. Such a scenario would be singularly unsatisfying, to say the least. And this leads to a rather radical concept. The Universe is normally thought of as a vast collection of tangible objects. Instead, perhaps we should consider it to be a "process". Processes can have beginnings and endings, and they follow rules that dictate behaviour, just as objects do. At the end of the process, space created and grown within the Universe would collapse, thus returning all matter-energy to the state-of-origin to become available for the next universal cycle. There is no knowing how many concurrent universes the Uberverse might be able to support, but even if it is limited to one, it might potentially enable a cycle that could repeat for eternity.

One of the major problems with the standard model of our Universe is the requirement for the density of matter-energy to be near-infinite at the point of origin.

Density is a property assignable to matter-energy, and is a direct consequence of the associated spatial framework. In the Uberverse, density may not be of concern, or even exist, because the spatial framework could be entirely different and unrelated to anything we might recognize. The possibility that density as a property could be quite different before the Big Bang compared with afterwards, suggests some intriguing modelling scenarios. When the spatial framework is created inside our Universe, the concentration of matter-energy becomes identifiable and measurable as density. One consequence of super-high density is 'pressure', which has the innate ability to drive matter-energy to regions of lower pressure. Might it be possible that the pressure derived from the spontaneous creation of space around a quantity of matter-energy is sufficient to explain the expansion of our Universe?

The mind-blowing density of matter-energy at the start of the Universe is not the only major cause for conceptual perplexity. There is also the challenge of explaining how an expanding boundary of the Universe might butt up against a Uberverse in which space could be relatively uniform and unchanging. Could we logically expect that the background 'emptiness' into which the Universe expands is identical to the environment that existed prior to the Big Bang? The only correct answer is, "It's complicated". The balloon analogy is useful when attempting to visualize the expanding space within our Universe. When deflated, conditions inside and outside the balloon would be similar. The Quark-Gluon Plasma believed to exist during the earliest moments of the Universe may have closely resembled conditions outside the Universe just prior to the Big Bang. As the Universe expands, conditions on either side of the boundary become quite different, with the internals of the Universe seemingly diluted. Because the space-time of our Universe is unlikely to extend influence beyond its boundaries, the Universe might be viewed externally as an intense congregation of non-expanding matter-energy. How might the expanding space in our Universe interface with that of the Uberverse where the spatial framework could possibly be entirely different?

If conditions in the Uberverse at the expanding boundary of our Universe are the same as those prior to the Big Bang, could we expect to find evidence to support this idea? Probably not, and for several reasons. The Universe has been expanding at an increasing rate for billions of years, and any hypothetical boundary is now so far away and travelling so fast, that light, or any other detectable radiation, is unable to reach us. Even if light from the boundary could reach Earth, it is unlikely to convey useful information. Our view of a segment of the ever-expanding boundary would be akin to continuously ramping up the zoom-level of a digital image until all that is visible is the space between the pixels. In addition, we have no clues to indicate the visibility of entities and processes at the boundary. They well might be classified with the mysterious Dark Matter and Dark Energy.

Conceptualizing extremes of the unknown is never easy. When considering how the Big Bang might have been triggered, the distribution of matter-energy within the Uberverse would seem to be of fundamental importance. Looking at cosmic phenomena we're familiar with, self-ignition could have occurred due to the intense compression of a critical amount of matter-energy accumulating in one place, much the same as the thermonuclear reaction initiates in stars. Alternatively, bundles of matter-energy sufficient to create a universe could be in abundance, and rely upon some mysterious external trigger to start the process. Since time has little relevance in this environment, the trigger might be explosive and abrupt, or it might slowly build up to some point of instability that produces the eruption.

Vast amounts of historical data are being collected from the heavens as the Hubble and Webb space telescopes probe ever deeper. No matter how much data we collect during the remaining period of human civilization, it will never be sufficient to construct a complete picture of the cosmic puzzle. Some aspects of this picture require postulation and speculation that can never be empirically validated. The inability to monitor cosmic events from a ringside seat is not actually a bad thing. If it were possible, sitting too close to events of cataclysmic destruction would likely have a detrimental effect on the chances of survival!

So what clues might exist inside our Universe to help build a credible model of the environment that preceded the Big Bang? A logical place to start is to search for extreme concentrations of matter. The Universe is studded with Black Holes, some supermassive, but they seem unwilling to give up their secrets without a fight. Whether or not Black Holes can grow to critical mass sufficient to trigger the birth of a new universe, may never be known. Maybe all Black Holes eventually amalgamate to create a Big Bang scenario. What a universe expanding inside our own might look like, is anybody's guess. Apart from super-high density, we have no knowledge about any other ingredients in the mix.

But all is not lost – there is another card yet to be played. Technological advances have extended powers of detection to the point where it seems to cover the full spectrum of electromagnetic radiation. And yet we are still unable to directly detect dark matter. This implies that the science of detection still has a distance to run, potentially even involving 'new' science. Here might be the breakthrough opportunity to discover processes related to Dark Matter, and, if there is one, the relationship to Black Holes. That may well be the closest we can ever get to finding empirical evidence for the processes that might have been involved in universal creation.

Most discussions involving the Big Bang revolve around it being the centre of creation of our space-time framework. That's fine when dealing with matter and energy. Where the concept comes unstuck is in trying to imagine a mechanism that supports the instantaneous creation of an unimaginably-large number of rules to dictate behaviour of all matter-energy in the newborn universe. That's a much harder issue, particularly in light of the fact that these rules have all been immutable since they were first inaugurated, before our galaxy existed, before humans were ever thought of.

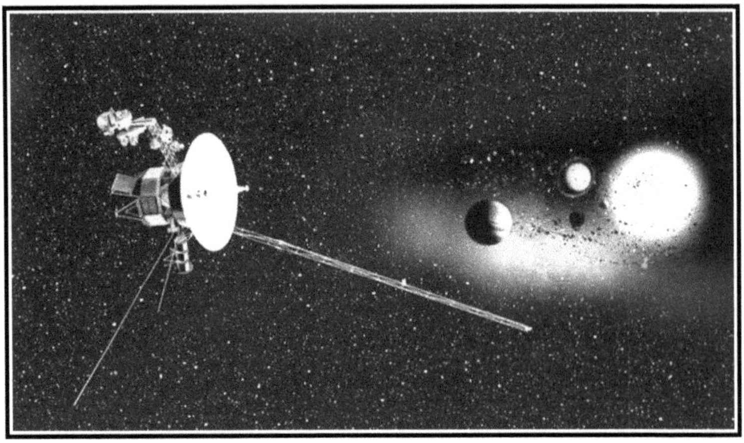

12. NO BEGINNING AND NO END

"There are more things in Heaven and Earth, Horatio, than are dreamt of in your philosophy" --
Shakespeare: Hamlet; Act 1

It is extremely difficult to imagine a point before the birth of our Universe where there were no rules, and then suddenly there were. Spontaneity of rule creation is a very difficult concept to explain in any meaningful way. However, if there is no necessity to establish a unique beginning for the complete set of Nature's rules, together with a corresponding ultimate end, then a paradox no longer exists.

When considering the situation at the beginning of our Universe, there are two very clear and unavoidable choices – either the rules were created at the moment of the Big Bang, or they were already in existence beforehand. Given the myriad of intimate interconnections between the rules, there would seem to be no middleground where they might have been modified at the last minute. Even if such modifications were possible, the rules governing modification must have been previously in place. For the instantaneous creation of the rules at a particular moment, there needs to be a superset of rules that control the processes responsible for generating the new set. There seems to be no escaping the fact that a set of Rules existed in some form or other prior to the creation of our Universe.

Our Universe could be seen as a container for all matter-energy within. The rules of nature permeate through the entire space of our Universe, although they are not restricted by any concept of containment. Similarly, the rules that initiated the Big Bang are not necessarily contained outside our Universe, or restricted from operating within it. However, within our Universe we may never be able to detect behaviour resulting from the application of those rules. The extreme environmental conditions required for these higher-level rules to operate only existed around the time of the Big Bang. The rules of nature may not recognize any artificial boundaries we have defined to contain our Universe.

At the smaller end of the scale, humans from birth are conditioned for death. An individual may witness the beginning and associated end of many things during their lifetime. We experience the idea of a beginning and an end with everything we know. It could be from the birth and death of our pet cat. It could be a realization of our own position on the mortality-scale. It could be the birth and death of our Sun. But it would be quite forgivable to extrapolate that concept of always having a beginning and end to apply to everything. Reaching as far back as possible, the Big Bang will be the earliest beginning we are ever likely to observe – we will never have any opportunity to look beyond that.

By virtue of their Earth-boundedness, humans have a very myopic view of existence. It might seem paradoxical to suggest that there is no beginning and no end, when everything within our range of perception appears to have a beginning and an end. There is no conflict if we consider evolving processes to be cyclical. There is a cycle of biological life; a cycle involving formation and destruction of planets; a cycle involving the birth and death of stars. It is also a possibility that universes like our own construct and destruct in cycles. There is a beginning and an end for each cycle, but no end to the cyclical process. The only constant throughout all eternity may well be the set of immutable Rules that govern existence.

If we subscribe to the idea that remnants from the end of our Universe could be recycled for use in future universes, it has important implications regarding the rules of nature. Recycling is a repetitive process within a consistent set of operating rules. And the implication? The Rules of Nature we observe within our Universe's space-time framework should apply equally outside that framework.

To comply with the notion that all physical manifestations exist on a timeline between start and end points, it is convenient to relate everything in our Universe back to a single point of origin – the Big Bang. However, there is a real problem when mapping things from the unknown onto the canvas of everyday experience. How can we be sure which conceptualization of 'beginning' best represents reality? For an individual human, the beginning is outside their control, and the time span to the end is strictly limited. They have no influence on beginnings and endings in the bigger picture. That said, it would be unfair to fail to acknowledge that there have certainly been some exceptional individuals whose influence on human civilization lasted a lot longer than they themselves did.

Historical records accessible by humans go back billions of years, but fall well short of providing any explicit detail about the Big Bang event at the start of our Universe. Because we will never have evidence to the contrary, it might seem reasonable to assume that all rules for existence were created at the time of the Big Bang. Would it be valid to also assume that the Big Bang was the start of EVERYTHING, just because the bigger picture is obscured? If everything else seems to have a beginning and ending, contained within other beginnings and endings, why should it be any different for the particular universe we just happen to be part of?

The idea that there is no beginning and no end, ties neatly with an hypothesis of cosmic continuity. There are many current theories that attempt to explain how our Universe might end, all of them equally unsupported by empirical evidence. Extrapolating from historical records of observation that form the basis of our cosmic understanding, it would not seem unreasonable to expect that any matter remaining after the end of our Universe might somehow be recycled for use in the creation of future universes. It doesn't really matter if the visible components of our Universe appear to disappear. If at the end of the game, all matter is eventually returned to the eternal background from whence it came, then recycling can go on indefinitely. There is a very satisfying continuity implied in this model.

Empirical evidence tells us that all the rules of Nature are immutable. So if we can willingly accept that these rules have been immutable since the start of our Universe, it's not such a giant leap to believe that this same set of rules might have also existed prior to the Big Bang. And that raises the prospect of continuity for rules that will apply for eternity.

Although it might appear that the rules of nature have been written to take account of every possible eventuality, nowhere is there evidence that any of these rules contain time-critical elements or dictate a sequence of application. For biological life, the opposite is true. Everything related to the evolutionary processes that ultimately lead to complex lifeforms, depends upon a sequence of events. These events can sometimes require extremely long periods to play out. The timing of any such development is purely a function of the environmental conditions, and not dependent on specifications in the generic rules for biology.

The window through which humans view everything is so frustratingly limited that we will never be able to identify with certainty any cyclical process of cosmic significance in which we might be involved. The short time-span allotted for human existence forces us to focus closely on the observable world and rely heavily on extrapolation to predict the future. And in that observable world, absolutely everything appears to have a 'use-by date'. This has fostered a common belief that when our Universe ends, it will be the end of everything. A major theme of this book has been to challenge that notion, and question why, after the establishment of such a magnificently beautiful set of rules defining existence, all would be abandoned just because one Universe has run its course. There is comfort to be had in the idea of continuity, and although this book takes the concept in a fairly philosophical context, this is also the approach that drives various religions. There are many things that will forever remain beyond our level of connection and comprehension.

So what is the detriment to science when we restrict our thinking to the hypothesis that there must always be a beginning and an end? A serious handicap is that we might be excluding examination of possibilities that exist outside such an hypothesis. If we can accept the theory of eternal continuity, we're free to examine a much bigger picture than just that of our own Universe. Undoubtedly this will remain forever in the realms of philosophy, but a huge part of human endeavour is devoted to philosophizing.

One of the doors opened by continuity theory invites us to contemplate what might happen at the end of our Universe. There seems to be no contradiction in believing our Universe has a beginning and an end, because the suggestion is that it is part of a bigger cycle where beginnings and endings happen all over the shop. So, as noted earlier, an exciting, recently-confirmed discovery from Cosmology is that our Universe is expanding at an ever-increasing rate (ref: **Expanding Universe**). This doesn't fit comfortably with the idea that it might be part of the cycle. It may well be that the end of our cycle results in oblivion for all the bits. How that ending might participate in the beginning of another universal cycle is something for examination in another place.

There are lots of theories about how the Universe might end, some more depressing than others. Getting back to this problem of what happened at the beginning, suppose that there is some entity above us who was responsible for creating all these complex rules. Then we're forced back even further to question where that entity might have had its beginnings. Working with the hypothesis that there must always be a beginning and end, really becomes too difficult to conceptualize. The alternative to this is to accept that, in the big picture, there was never a beginning and there will never be an ending. That greatly simplifies the issue of how it all started – it didn't have a beginning, it always was. Likewise, there would be no reason to anticipate an ultimate end.

All the mysterious elements are in the mix to allow modelling that would accommodate a turnaround in the rate of expansion of our Universe. We have these wondrously nebulous entities to temporarily plug any holes that become apparent when current theoretical models of the Universe are contradicted by the evidence – the Dark-Matter and Dark Energy. This presents an opportunity to investigate what mechanisms could be involved that might possibly put the brakes on the Universe's expansion to oblivion. As proposed in the previous chapter, the Universe might not need to incorporate a reversal to be cyclical – space within the Universe could collapse at some point, returning all contained matter-energy to its primal state.

If we break with the constraints imposed by beginnings and endings, we are free to search for physical evidence in support of cosmic continuity. One of the biggest challenges to scientific thinking is the notion that 'something can come from nothing'. The rules identified thus far in Physics and Thermodynamics lead to the conclusion that matter-energy is always preserved, and cannot be created or destroyed. And yet, that's one of the fundamental requirements if you subscribe to the idea of the Big Bang as a singularity without any context. It is outside the realm of this book to touch on the mind-bogglingly complex concept of multiple dimensions existing in space and time. Mathematical models have been developed that make this sound almost plausible.

If we can accept that all the matter-energy of our Universe was not created at the moment of the Big Bang, then it may be possible to consider it as merely a portal connecting to somewhere else. The origin of our

Universe could be a portal with stuff still pouring through. That theory may not be quite as crazy as it sounds – it would certainly provide an alternative explanation for the increasing rate of expansion. Perhaps the Big Bang might actually itself have continuity. The model wherein the Big Bang spontaneously came into existence at some point outside the Space-Time framework of our Universe, and then expanded solely due to explosive forces, does not fit the evidence. Maybe as Douglas Adams has suggested, we're all just part of a giant experiment (ref: **Hitchhikers Guide to the Galaxy**).

Once humans break the emotional bond to the notion that they are special, the question then arises as to who are beneficiaries of the existence of everything. A logical answer is that all life benefits from the opportunity to exist, no matter where in the Universe it may be. And that leads to another intriguing question – if the same philosophy of giving life a chance is the driver for all existence, it would seem strange if this was restricted to the single universe we happen to reside in. Would it be reasonable to expect the same rules and opportunities for life would apply within universes created outside our own?

One of the dangers in pursuing the really hard questions is the possibility of finding answers that are less than desirable or attractive. An inevitable conclusion from examining the big picture, is that humankind is truly insignificant. But does that really discount the exhilaration of achievement during the short time we have been permitted on this journey? Sometimes we may need to be reminded that on the voyage of discovery, the joy is in the journey, rather than in arrival at some ephemeral destination involving the understanding of everything.

In an earlier section, we made reference to the most significant question of all time – Why does anything exist at all? If we ever determine the answer, a follow-up question would be – Why should we expect our Universe to be the only one? Space and Time have a special meaning within the context of our Universe. Both are quantifiables of restricted applicability. Outside the framework of our Universe, it would be hard to imagine that Space and Time are restricted in some manner. The boundaries would presumably extend to infinity through all eternity. The intricate and magnificently beautiful set of rules driving our existence in this universe would seem to lack justification if they were destined for use only in a single implementation.

The more we appreciate the infiniteness of space and time, the better we will be able to understand the meaning of life. With no permanent record to show that humanity ever existed, and no chance of influencing the distant future in any way, the significance of life must be "in the moment". And a thought to go... Long after our solar system has been devoured by a Black Hole, those tiny Voyager spacecraft that headed off to interstellar space 50 years ago, may contain the only surviving evidence of human civilization. That lonely record gives some indication of the way we have interpreted the reasons and rules for existence.

The End (... or maybe NOT)

REFERENCE MATERIAL

Abiogenesis – How Did Life on Earth Begin
https://theconversation.com/did-life-evolve-more-than-once-researchers-are-closing-in-on-an-answer-205678

Animal Infanticide – Why Infanticide Can Benefit Animals
https://en.wikipedia.org/wiki/Infanticide_(zoology)

Anthropomorphism – Personification of Deities
https://www.britannica.com/EBchecked/topic/27536/anthropomorphism

Antimatter Gravitation – Experiment confirms Antimatter falls downwards
https://home.cern/news/press-release/physics/alpha-experiment-cern-observes-influence-gravity-antimatter

Creationism – All Life resulted from Divine Intervention
https://en.wikipedia.org/wiki/Creationism

Dark Energy Source – Scientists find connection with Black Holes
https://phys.org/news/2023-02-scientists-evidence-black-holes-source.html

Effects of Weightlessness
https://www.smithsonianmag.com/smart-news/why-astronauts-have-weaker-immune-systems-in-space-180982423/

Emotion in Animals
https://en.wikipedia.org/wiki/Emotion_in_animals

Expansion of the Universe – Inflationary Epoch
https://en.wikipedia.org/wiki/Inflationary_epoch

Expanding Universe – Speed of Universe's Expansion Measured Better Than Ever
https://www.space.com/17884-universe-expansion-speed-hubble-constant.html

Extremophiles – Extremes of Life on Earth
https://www.bbc.com/future/article/20140303-last-place-on-earth-without-life

Genesis – God created Man as a Special Species
https://en.wikipedia.org/wiki/Genesis_creation_narrative

Geostationary Orbit – Arthur C. Clarke
https://en.wikipedia.org/wiki/Geostationary_orbit

Graphene – A New Form of Carbon
https://en.wikipedia.org/wiki/Graphene

History of Earth – The biggest turning points in Earth's history
https://www.bbc.com/earth/bespoke/story/20150123-earths-25-biggest-turning-points/index.html

Hitchhikers Guide to the Galaxy – Douglas Adams
https://www.britannica.com/topic/The-Hitchhikers-Guide-to-the-Galaxy-novel-by-Adams

Human Evolution – How have we changed since our species first appeared?
https://australianmuseum.net.au/How-have-we-changed-since-our-species-first-appeared

ITER Fusion Reactor – World's Largest Nuclear-Fusion Experiment
https://www.iter.org/proj/inafewlines

Mapping the Universe – Mapping the Universe and measuring expansion rate
https://newscenter.lbl.gov/2024/04/04/desi-first-results-make-most-precise-measurement-of-expanding-universe/

Neurology of Space Flight – Space flight effects on human nervous system
https://www.sciencedirect.com/science/article/pii/S2214552422000694

Pantheism – A Nature-Based Religion
https://www.pantheism.net/

Quantum Computing – What are they good for?
https://www.nature.com/articles/d41586-023-01692-9

Quantum Entanglement – Advances in Quantum Teleportation
https://www.jpl.nasa.gov/news/news.php?feature=4384

Quark-Gluon Plasma – Large Hadron Collider – Experimentally Creating QGP
https://www.popsci.com/science/large-hadron-collider-quark-gluon-plasma/

Situs Inversus Totalis – Reversal of Internal Body Organs
https://www.ncbi.nlm.nih.gov/pmc/articles/PMC8901252/

Spacecraft Voyager I and Voyager II still functional after nearly 50 years
https://blogs.nasa.gov/sunspot/2023/12/12/engineers-working-to-resolve-issue-with-voyager-1-computer/

Steady-State Cosmology – Jeans, Bondi, Gold, Hoyle, Burbidge, and Narlikar
https://handwiki.org/wiki/Astronomy:Steady_state_model

Stem Cell Niche – Micro-environment for stem cells
https://en.wikipedia.org/wiki/Stem_cell_niche

Synthetic Cell Growth – Reproducing Life in the Laboratory
https://scitechdaily.com/artificial-life-forged-in-a-lab-scientists-create-synthetic-cell-that-grows-and-divides-normally/

Telomeres & Immortality – Can Science Make Us Immortal
https://www.viewzone.com/aging.html

Testing Relativity Theory – Einstein's Theories
https://scitechdaily.com/einstein-proven-right-yet-again-theory-of-general-relativity-passes-a-range-of-precise-tests/

The Cosmic Calendar – Carl Sagan
https://en.wikipedia.org/wiki/Cosmic_Calendar

The Gaia Hypothesis – James Lovelock, Lynn Margulis
https://www.gaiatheory.org/overview/

The God of the Gaps
https://en.wikipedia.org/wiki/God_of_the_gaps

The Higgs Boson – Experiments at CERN
https://home.web.cern.ch/topics/higgs-boson

The Rosetta Mission – Evidence of Organic Material Originating Beyond Earth
https://spaceflightnow.com/2014/11/18/philae-finds-comet-harbors-organics/

Thermal-Vent Creatures
https://seawifs.gsfc.nasa.gov/OCEAN_PLANET/HTML/ps_vents.html

Turritopsis Dohrnii – The Immortal Jellyfish
https://en.wikipedia.org/wiki/Turritopsis_dohrnii

Understanding of Aerofoil Lift
https://aia.springeropen.com/articles/10.1186/s42774-021-00089-4

Unsustainable Consumption – Resources running out; Urgent action required
https://www.weforum.org/agenda/2024/03/sustainable-resource-consumption-urgent-un/

ABOUT THE AUTHOR

John was raised in country Victoria, Australia, where an uncluttered lifestyle provided ample opportunity for hands-on exposure to the wonders of nature. At the University of Melbourne he took out an honours degree in Science, thus qualifying as a Particle Physicist. Although he was well-qualified and enthusiastic, it soon became apparent that nuclear-phobic Australia offered quite limited career prospects in that field. He then completed a second degree in Electrical Engineering.

For several years John worked with Schlumberger as an oil engineer, before retiring to coastal Queensland where he established a small electronics business. It was not commonly known, but this business was purely a front to facilitate his desire to invent things. His proudest achievement was possibly the Computaphon, the world´s first electronic phone.

None of John´s inventions was ever commercialized as he expressed little interest beyond building prototypes and proof-of-concept. He later gravitated to software development, a pursuit which continues to this day. However, none of these ´day jobs´ managed to overshadow his fundamental love –- Cosmology.

Alphabetical Index

Abiogenesis..65, 180
Aerofoil...72, 184
Algorithms...........................73 f., 104, 126, 129 f.
Alien...29, 57, 88, 97 f., 100 f.
Ancestor................3, 7, 31, 35, 39, 41, 43, 70, 149
Anomalies...110, 114
Anthropomorphism....................................54, 180
Antigravity...115
Antimatter..114 f., 180
Artificial Intelligence..................76, 104 f., 130 f.
Astronauts...98, 147, 180
Astrophysics..95, 159
Atmosphere.................37, 45, 76, 84, 87, 100, 139
Autonomous...105, 127 f.
Bilateral...40, 87, 144
Binary Computers......................................129 f.
Biosphere..53
Bisymmetrical...148
Black Holes......................101, 116, 165, 178, 180
Chaos...................................vi, 40, 48, 74, 90, 121, 157
Chicken...47, 128
Chromosomes..21
Colonization.......................................96 f., 99, 101
Continuity...58, 172 ff., 176 f.
Conundrum...115, 148
Cosmic Inflation..118 ff.
Cosmology.............iii, 48, 53, 58, 112, 115, 175, 183, 185
Cosmos...........................8 f., 21, 48, 54, 87, 110 f., 120
Creationism..44, 59, 180

Creator..51, 58 ff.
Dangers................................2, 56, 88, 104, 130, 177
Dark Matter........................90, 113 ff., 119, 163, 165
Darwin Charles.........................39, 47, 59, 143, 148
Density............................31, 69, 119, 158, 161 f., 165
Destiny...2, 11, 85, 87
Deuterium...90
Dinosaurs...................................7, 21 f., 30, 48, 86 ff.
Doomsday..................................36, 60, 76, 131
Douglas Adams.......................................177, 181
Ecosystems......................85, 99, 105, 139, 141
Einstein Albert.............68, 81, 88, 102, 109, 112, 120, 183
Electromagnetism...........37, 83 f., 113 f., 121, 138, 140, 165
Electromechanical...131
Electrostatic Force..13
Entanglement...........................89, 93 f., 104, 182
Equilibrium..................5, 10, 35, 60, 64, 76, 100, 121
Eternal...81, 173 f.
Event Horizon..vi
Extinction................................24, 31, 53, 75 f., 85, 88
Extraterrestrial Life..46
Extremophiles...151, 181
Galaxies............8, 13, 27, 47, 65, 101, 116, 141, 166, 177, 181
Genes........................24, 38, 40, 56, 65, 149, 158, 180 f.
Geostationary Orbit..................................v, 181, 183
Global Warming..5, 17, 76
God.............................4, 51, 54 ff., 58, 61, 181, 183
Goldilocks Zone...65, 139
Graphene..14, 181
Helium..27
Higgs Boson..12, 114, 183

Homo Sapiens	33
Hoyle Fred	112, 183
Hubble	164, 181
Human Intelligence	87, 103 f., 128, 130, 147 f.
Humanoid	100, 126 f., 145
Hydrogen	27, 90
Hypothesis	4, 7, 104, 172, 174 f., 183
Immortality	23, 183
Implants	133 f.
Inbreeding	40, 99
Industrialization	5 f., 142
Inertia	30, 94 f.
Infanticide	4, 180
Instability	v, 79, 82, 89, 164
Instinct	3, 23, 71
Intangibles	66
Isotopes	90
Jeans James	112, 183
Lego	27 ff.
Lifecycles	141
Lifespan	11, 23, 35, 75, 80, 105, 146
Magnetism	37, 83 f., 97, 104, 114, 138, 140, 165
Mapping the Universe	115, 171, 182
Metabolism	98, 102
Meteorite	76, 86 f.
Microbes	85, 98
Microgravity	148
Microorganisms	36, 98, 139
Molecular	33, 70, 75, 125, 137
Monotheism	51
Mutations	24, 38, 40, 83 ff., 89, 96, 143 f., 149

Newtonian Mechanics..67 f.
Nuclear Fusion....................................40, 112, 138, 181
Nuclear Reactor...................................112, 138 f., 181
Nucleons................13, 76, 90, 112, 118, 138 ff., 164, 181, 185
Overpopulation...76
Pantheism...56, 182
Paradigm Shift...130
Paradox...47, 169, 171
Paranormal..103 f.
Pi Meson...76
Planetary..52, 97, 141 f.
Plum Pudding..69
Polytheism..51
Portal..176 f.
Positrons..114
Probabilistic...............................60, 70, 75, 82, 87, 90
Procreation..3, 24, 39
Prosthetic..125, 128
Quark-Gluon Plasma...............................118 f., 162, 182
Radiation......................37 f., 83 f., 97, 114, 139 f., 163, 165
Radioactive..89 f.
Radioactive Decay...13, 89 f.
Recycling...141, 160, 171 ff.
Regeneration...24 f.
Relativity Theory..................................66, 68, 120, 183
Religion..................................51 f., 54 ff., 60, 174, 182
Replication...8, 23, 84
Repulsion..13
Robots...105, 125 ff.
Rosetta Mission..47, 183
Sagan Carl...52, 183

Shakespeare William	45, 88, 169
Shields	95, 97, 139 f., 150
Simulation	97 f., 118, 129
Singularities	57, 161, 176
Situs Inversus Totalis	149, 182
Solar System	8, 17, 47, 65, 80, 178
Space Shuttle	96
Spacecraft	80, 98, 178, 182
Superbugs	76
Sustainability	7, 24, 76, 86, 101
Symbiotic Relationship	36, 98
Symmetry	40 f., 74, 87, 144, 148
Telepathy	103 f.
Telescopes	164
Telomeres	23 f., 183
Thermal Vents	37, 84, 184
Thermodynamics	176
Time Dilation	120, 159
Timelines	2, 17, 76, 85, 137, 171
Transistors	44, 105, 122
Tritium	90
Turritopsis Dohrnii	25, 184
Uberverse	155 ff.
Unsustainable	6, 146, 184
Uranium	90
Voyager Spacecraft	80, 178, 182
Weightlessness	98, 180

Milton Keynes UK
Ingram Content Group UK Ltd.
UKHW022145111124
451073UK00007B/198